TOXICOLOGIC ASSESSMENT OF THE ARMY'S ZINC CADMIUM SULFIDE DISPERSION TESTS

SUBCOMMITTEE ON ZINC CADMIUM SULFIDE

COMMITTEE ON TOXICOLOGY

BOARD ON ENVIRONMENTAL STUDIES AND TOXICOLOGY

COMMISSION ON LIFE SCIENCES

NATIONAL RESEARCH COUNCIL

NATIONAL ACADEMY PRESS
WASHINGTON, D.C., 1997

NATIONAL ACADEMY PRESS   2101 Constitution Avenue, NW   Washington, DC 20418

NOTICE: The project that is the subject of this report was approved by the Governing Board of the National Research Council, whose members are drawn from the councils of the National Academy of Sciences, the National Academy of Engineering, and the Institute of Medicine. The members of the committee responsible for the report were chosen for their special competences and with regard for appropriate balance.

This report has been reviewed by a group other than the authors according to procedures approved by a Report Review Committee consisting of members of the National Academy of Sciences, the National Academy of Engineering, and the Institute of Medicine.

The project was supported by contract DAMD 17-89-C-9086 between the National Academy of Sciences and the U.S. Army. Any opinions, findings, conclusions, or recommendations expressed in this publication are those of the author(s) and do not necessarily reflect the view of the organizations or agencies that provided support for this project.

Library of Congress Catalog Card Number 97-67216
International Standard Book Number 0-309-05783-3

Additional copies of this report are available from:

National Academy Press
2101 Constitution Ave., NW
Box 285
Washington, DC 20055
800-624-6242 or 202-334-3313 (in the Washington Metropolitan Area)
http://www.nap.edu

Copyright 1997 by the National Academy of Sciences. All rights reserved.

Printed in the United States of America.

SUBCOMMITTEE ON ZINC CADMIUM SULFIDE

ROGENE F. HENDERSON *(Chair)*, Lovelace Biomedical and Environmental Research Institute, Albuquerque, New Mexico
GERMAINE M. BUCK, State University of New York at Buffalo, New York
JOHN E. CONNETT, University of Minnesota, Minneapolis, Minnesota
ELAINE FAUSTMAN, University of Washington, Seattle, Washington
CHARLES E. FEIGLEY, University of South Carolina, Columbia, South Carolina
DONALD E. GARDNER, Inhalation Toxicology Associates, Raleigh, North Carolina
DAVID W. GAYLOR, U.S. Food and Drug Administration, Jefferson, Arkansas
ROBERT A. GOYER, National Institute of Environmental Health Sciences, Research Triangle Park, North Carolina
LOREN D. KOLLER, Oregon State University, Corvallis, Oregon
STEPHEN U. LESTER, Citizens Clearing House for Hazardous Waste, Falls Church, Virginia
THOMAS E. MCKONE, University of California, Berkeley, California
MICHAEL J. THUN, American Cancer Society, Atlanta, Georgia
BAILUS WALKER, JR., Howard University, Washington, D.C.
SUSAN D. WILTSHIRE, JK Research Associates, Inc., Hamilton, Massachusetts
HANSPETER R. WITSCHI, University of California, Davis, California

Sponsor: U.S. Army

Staff

KULBIR S. BAKSHI, Project Director
DIANE J. MUNDT, Senior Program Officer
ERIN M. BELL, Research Associate
SUSAN N.J. PANG, Research Associate
RUTH E. CROSSGROVE, Staff Associate
NORMAN GROSSBLATT, Editor
LINDA LEONARD, Senior Project Assistant
LUCY V. FUSCO, Project Assistant
KATHRINE IVERSON, Information Specialist

COMMITTEE ON TOXICOLOGY

ROGENE F. HENDERSON *(Chair)*, Lovelace Biomedical and Environmental Research Institute, Albuquerque, New Mexico
DONALD E. GARDNER *(Vice-Chair)*, Inhalation Toxicology Associates, Raleigh, North Carolina
GERMAINE M. BUCK, State University of New York at Buffalo, New York
DEBORAH A. CORY-SLECHTA, University of Rochester, Rochester, New York
KEVIN E. DRISCOLL, Procter & Gamble Company, Cincinnati, Ohio
ELAINE M. FAUSTMAN, University of Washington, Seattle, Washington
CHARLES E. FEIGLEY, University of South Carolina, Columbia, South Carolina
DAVID W. GAYLOR, U.S. Food and Drug Administration, Jefferson, Arkansas
IAN A. GREAVES, University of Minnesota, Minneapolis, Minnesota
SIDNEY GREEN, Corning Hazleton, Inc., Vienna, Virginia
WILLIAM E. HALPERIN, National Institute for Occupational Safety and Health, Atlanta, Georgia
LOREN D. KOLLER, Oregon State University, Corvallis, Oregon
GEORGE B. KOELLE, University of Pennsylvania, Philadelphia, Pennsylvania
DANIEL KREWSKI, Health Canada, Ottawa, Ontario
THOMAS E. MCKONE, University of California, Berkeley, California
MICHELE A. MEDINSKY, Chemical Industry Institute of Toxicology, Research Triangle Park, North Carolina
JOHN L. O'DONOGHUE, Eastman Kodak Company, Rochester, New York
ROBERT SNYDER, Environmental and Occupational Health Sciences Institute, Piscataway, New Jersey
BERNARD M. WAGNER, Wagner Associates, Inc., Millburn, New Jersey
BAILUS WALKER JR., Howard University, Washington, D.C.
ANNETTA P. WATSON, Oak Ridge National Laboratory, Oak Ridge, Tennessee
HANSPETER R. WITSCHI, University of California, Davis, California
GAROLD S. YOST, University of Utah, Salt Lake City, Utah

Staff

KULBIR S. BAKSHI, Program Director
MARGARET E. MCVEY, Program Officer
SUSAN N.J. PANG, Research Associate
ABIGAIL STACK, Research Associate
RUTH E. CROSSGROVE, Editor
CATHERINE M. KUBIK, Senior Program Assistant
LINDA V. LEONARD, Senior Project Assistant
LUCY V. FUSCO, Project Assistant

BOARD ON ENVIRONMENTAL STUDIES AND TOXICOLOGY

PAUL G. RISSER *(Chair)*, Oregon State University, Corvallis, Oregon
MAY R. BERENBAUM, University of Illinois, Urbana, Illinois
EULA BINGHAM, University of Cincinnati, Cincinnati, Ohio
PAUL BUSCH, Malcolm Pirnie, Inc., White Plains, New York
EDWIN H. CLARK II, Clean Sites, Inc., Alexandria, Virginia
ELLIS COWLING, North Carolina State University, Raleigh, North Carolina
GEORGE P. DASTON, The Procter & Gamble Co., Cincinnati, Ohio
PETER L. DEFUR, Virginia Commonwealth University, Richmond, Virginia
DAVID L. EATON, University of Washington, Seattle, Washington
DIANA FRECKMAN, Colorado State University, Ft. Collins, Colorado
ROBERT A. FROSCH, Harvard University, Cambridge, Massachusetts
DANIEL KREWSKI, Health & Welfare Canada, Ottawa, Ontario
RAYMOND C. LOEHR, The University of Texas, Austin, Texas
WARREN MUIR, Hampshire Research Institute, Alexandria, Virginia
GORDON ORIANS, University of Washington, Seattle, Washington
GEOFFREY PLACE, Hilton Head, South Carolina
BURTON H. SINGER, Princeton University, Princeton, New Jersey
MARGARET STRAND, Bayh, Connaughton and Malone, Washington, D.C.
BAILUS WALKER, JR., Howard University, Washington, D.C.
GERALD N. WOGAN, Massachusetts Institute of Technology, Cambridge, Massachusetts
TERRY F. YOSIE, E. Bruce Harrison Co., Washington, D.C.

Staff

JAMES J. REISA, Director
DAVID J. POLICANSKY, Associate Director and Program Director for Natural Resources and Applied Ecology
CAROL A. MACZKA, Program Director for Toxicology and Risk Assessment
LEE R. PAULSON, Program Director for Information Systems and Statistics
RAYMOND A. WASSEL, Program Director for Environmental Sciences and Engineering

COMMISSION ON LIFE SCIENCES

THOMAS D. POLLARD *(Chair)*, The Salk Institute, La Jolla, California
FREDERICK R. ANDERSON, Cadwalader, Wickersham & Taft, Washington, D.C.
JOHN C. BAILAR III, University of Chicago, Chicago, Illinois
PAUL BERG, Stanford University School of Medicine, Stanford, California
JOHN E. BURRIS, Marine Biological Laboratory, Woods Hole, Massachusetts
SHARON L. DUNWOODY, University of Wisconsin, Madison, Wisconsin
URSULA W. GOODENOUGH, Washington University, St. Louis, Missouri
HENRY W. HEIKKINEN, University of Northern Colorado, Greeley, Colorado
HANS J. KENDE, Michigan State University, East Lansing, Michigan
SUSAN E. LEEMAN, Boston University School of Medicine, Boston, Massachusetts
THOMAS E. LOVEJOY, Smithsonian Institution, Washington, D.C.
DONALD R. MATTISON, University of Pittsburgh, Pittsburgh, Pennsylvania
JOSEPH E. MURRAY, Wellesley Hills, Massachusetts
EDWARD E. PENHOET, Chiron Corporation, Emeryville, California
EMIL A. PFITZER, Research Institute for Fragrance Materials, Hackensack, New Jersey
MALCOLM C. PIKE, University of Southern California, Los Angeles, California
HENRY C. PITOT III, University of Wisconsin, Madison, Wisconsin
JONATHAN M. SAMET, The Johns Hopkins University, Baltimore, Maryland
CHARLES F. STEVENS, The Salk Institute, La Jolla, California
JOHN L. VANDEBERG, Southwest Foundation for Biomedical Research, San Antonio, Texas

PAUL GILMAN, Executive Director

OTHER RECENT REPORTS OF THE BOARD ON ENVIRONMENTAL STUDIES AND TOXICOLOGY

Carcinogens and Anticarcinogens in the Human Diet: A Comparison of Naturally Occurring and Synthetic Substances (1996)
Upstream: Salmon and Society in the Pacific Northwest (1996)
Science and the Endangered Species Act (1995)
Wetlands: Characteristics and Boundaries (1995)
Biologic Markers (Urinary Toxicology (1995), Immunotoxicology (1992), Environmental Neurotoxicology (1992), Pulmonary Toxicology (1989), Reproductive Toxicology (1989))
Review of EPA's Environmental Monitoring and Assessment Program (three reports, 1994-1995)
Science and Judgment in Risk Assessment (1994)
Ranking Hazardous Waste Sites for Remedial Action (1994)
Pesticides in the Diets of Infants and Children (1993)
Issues in Risk Assessment (1993)
Setting Priorities for Land Conservation (1993)
Protecting Visibility in National Parks and Wilderness Areas (1993)
Dolphins and the Tuna Industry (1992)
Hazardous Materials on the Public Lands (1992)
Science and the National Parks (1992)
Animals as Sentinels of Environmental Health Hazards (1991)
Assessment of the U.S. Outer Continental Shelf Environmental Studies Program, Volumes I-IV
(1991-1993)
Human Exposure Assessment for Airborne Pollutants (1991)
Monitoring Human Tissues for Toxic Substances (1991)
Rethinking the Ozone Problem in Urban and Regional Air Pollution (1991)
Decline of the Sea Turtles (1990)
Tracking Toxic Substances at Industrial Facilities (1990)

Copies of these reports may be ordered from
the National Academy Press
(800) 624-6242
(202) 334-3313

OTHER RECENT REPORTS OF THE COMMITTEE ON TOXICOLOGY

Toxicity of Military Smokes and Obscurants, Volume 1 (1997)
Toxicity of Alternatives to Chlorofluorocarbons: HFC-134a and HCFC-123 (1996)
Permissible Exposure Levels for Selected Military Fuel Vapors (1996)
Spacecraft Maximum Allowable Concentrations for Selected Airborne Contaminants, Volume 1 (1994), Volume 2 (1996), and Volume 3 (1996)
Nitrate and Nitrite in Drinking Water (1995)
Guidelines for Chemical Warfare Agents in Military Field Drinking Water (1995)
Review of the U.S. Naval Medical Research Institute's Toxicology Program (1994)
Health Effects of Permethrin-Impregnated Army Battle-Dress Uniforms (1994)
Health Effects of Ingested Fluoride (1993)
Guidelines for Developing Community Emergency Exposure Levels for Hazardous Substances (1993)
Guidelines for Developing Spacecraft Maximum Allowable Concentrations for Space Station Contaminants (1992)
Review of the U.S. Army Environmental Hygiene Agency Toxicology Division (1991)
Permissible Exposure Levels and Emergency Exposure Guidance Levels for Selected Airborne Contaminants (1991)

The National Academy of Sciences is a private, nonprofit, self-perpetuating society of distinguished scholars engaged in scientific and engineering research, dedicated to the furtherance of science and technology and to their use for the general welfare. Upon the authority of the charter granted to it by the Congress in 1863, the Academy has a mandate that requires it to advise the federal government on scientific and technical matters. Dr. Bruce Alberts is president of the National Academy of Sciences.

The National Academy of Engineering was established in 1964, under the charter of the National Academy of Sciences, as a parallel organization of outstanding engineers. It is autonomous in its administration and in the selection of its members, sharing with the National Academy of Sciences the responsibility for advising the federal government. The National Academy of Engineering also sponsors engineering programs aimed at meeting national needs, encourages education and research, and recognizes the superior achievements of engineers. Dr. William A. Wulf is president of the National Academy of Engineering.

The Institute of Medicine was established in 1970 by the National Academy of Sciences to secure the services of eminent members of appropriate professions in the examination of policy matters pertaining to the health of the public. The Institute acts under the responsibility given to the National Academy of Sciences by its congressional charter to be an adviser to the federal government and, upon its own initiative, to identify issues of medical care, research, and education. Dr. Kenneth I. Shine is president of the Institute of Medicine.

The National Research Council was organized by the National Academy of Sciences in 1916 to associate the broad community of science and technology with the Academy's purposes of furthering knowledge and advising the federal government. Functioning in accordance with general policies determined by the Academy, the Council has become the principal operating agency of both the National Academy of Sciences and the National Academy of Engineering in providing services to the government, the public, and the scientific and engineering communities. The Council is administered jointly by both Academies and the Institute of Medicine. Dr. Bruce Alberts and Dr. William A. Wulf are chairman and vice chairman, respectively, of the National Research Council.

Preface

During the 1950s and 1960s, the U.S. Army conducted dispersion tests using particles of zinc cadmium sulfide (ZnCdS) as a nonbiologic simulant of biologic-warfare agents in a number of urban and rural locations in the United States and Canada. This report, by the Subcommittee on Zinc Cadmium Sulfide of the National Research Council's Committee on Toxicology, is intended to assist the Army and the U.S. Congress in their efforts to determine whether exposure to ZnCdS particles adversely affected the health of persons living in the areas where the dispersion tests were conducted. The report independently reviews the available toxicity data on ZnCdS and its components cadmium and zinc, assesses human exposures to ZnCdS, and characterizes the risk to people exposed to it through the Army's dispersion tests.

The subcommittee was greatly assisted by several persons who provided information on the Army's ZnCdS dispersion tests and toxicity data on ZnCdS and its components. The subcommittee gratefully wishes to acknowledge William Barnett, Amy Birks, Leslie Burger, John Doesberg, Dennis Druck, Frederick Erdtman, Robert M. Gum, Bernard Ingold, Jeffrey Kirkpatrick, Francis O'Donnell, Forrest Oliverson, John Riggs, and Carmen J. Spencer—all of the U.S. Army—for their interest in and support of the project, as well as Amy S. Adair, Kenneth R. Boley, Monica Chavez, Cathy M. Collier, Marjorie A. Duske, Michael Erlandson, Christine Hawk, Thomas C. Keller, John L. Less, Derek D. Lick, Dobie O. McArthur, and Jennifer M. Tisdale—of the staff of the U.S. Congress. We are grateful to Edmund Crouch of Cambridge Environmental, Inc., for reviewing the Army's data on the ZnCdS dispersion tests and for estimating the doses and concentrations of ZnCdS particles. We are also grateful to Sheila Fabiano (USR Optronix), Terry Gordon (New York University Medical Center), Günter Oberdörster (University of Rochester), and Bruce Parkinson (University of Colorado), for making presentations or providing material to the subcommittee. The subcommittee also wishes to thank the several hundred persons who took the time and made the effort to meet with or present material to the subcommittee at the public meetings.

As chair of the subcommittee, I am grateful for the assistance of the National Research Council staff in the preparation of the report. Staff members who contributed to this effort are Paul Gilman, executive director of the Commission on Life Sciences; James J. Reisa, director of the Board on Environmental Studies and Toxicology; Carol A. Maczka, program director for toxicology and risk assessment; Diane J. Mundt and Erin M. Bell of the Medical Follow-up Agency, Institute of Medicine; Jamie E. Young, Board on Environmental Studies and Toxicology; Norman Grossblatt, editor; Susan N.J. Pang, research associate; Ruth E. Crossgrove, staff associate; and Lucy V. Fusco, project assistant. I especially wish to recognize the major contributions of the project director, Kulbir S. Bakshi, who exhaustively studied the literature and drafted several sections of the report. He worked tirelessly to obtain information, and he organized the study plan, the subcommittee and public meetings, the special presentations, and this final report.

Finally, I would like to thank all the members of the subcommittee (who worked without compensation in public service) for their expertise and dedicated effort throughout the development of this report. The members of the subcommittee represented an unusually diverse set of disciplines, from laboratory scientists to epidemiologists to risk-communication experts. Despite this diversity, which was essential for consideration of the many issues involved in the problem we were addressing, the members worked together in a highly effective manner. Discussions were open, honest, and vigorous. For the untiring efforts of these dedicated people, I am sincerely grateful!

                                        Rogene F. Henderson, PhD
                                        Chair, Subcommittee on Zinc Cadmium Sulfide
                                        and *Chair*, Committee on Toxicology

CONTENTS

	SUMMARY	1
1	INTRODUCTION	17
	ZnCdS Dispersion Tests	18
	Public Concern in Response to ZnCdS Dispersion Tests	19
	Tasks of the Subcommittee	20
	Sources of Data	22
	Structure of the Report	24
2	INPUT FROM THE PUBLIC	25
	Nature of Comments	26
	Specific Health Problems	29
	Conclusions	30
3	TOXICITY AND RELATED DATA ON ZINC CADMIUM SULFIDE	32
	Physical and Chemical Properties	32
	Toxicokinetics and Bioavailability of ZnCdS: Availability of Cadmium from ZnCdS	35
	Toxicity	35
	Implications of Variable Composition	37
	Conclusions and Recommendations	37
4	TOXICITY AND RELATED DATA ON SELECTED CADMIUM COMPOUNDS	39
	Physical and Chemical Properties of Cadmium Compounds	40

	Toxicokinetics of Cadmium Compounds	42
	Toxicity of Cadmium Compounds	43
	Carcinogenicity of Cadmium Compounds	51
	Conclusions	55
5	**EXPOSURE ASSESSMENT**	57
	Zinc Cadmium Sulfide	57
	Cadmium	60
	Conclusions	65
6	**RISK CHARACTERIZATION OF EXPOSURES TO ZINC CADMIUM SULFIDE**	67
	Risk Assessment for Noncancer Health Effects	68
	Risk Assessment for Cancer	72
	Conclusions	79
7	**SCIENTIFIC FEASIBILITY OF EPIDEMIOLOGIC STUDY**	82
	Nature of Epidemiologic Investigations	83
	Key Methodologic Issues	84
	Types of Epidemiologic Studies	89
	Conclusions	91
8	**CONCLUSIONS AND RECOMMENDATIONS**	94
	Input from the Public	94
	Toxicity and Related Data on Zinc Cadmium Sulfide	95
	Toxicity and Related Data on Selected Cadmium Compounds	95
	Exposure Assessment	96
	Risk Assessment of ZnCdS Exposures	97
	Feasibility of Epidemiologic Study	98
	Recommendations	98
	REFERENCES	100
	APPENDIX A: Historical Background of the U.S. Biological Warfare Program	112
	APPENDIX B: Summary of Doses and Concentrations of Zinc Cadmium Sulfide Particles from the Army's Dispersion Tests	122
	APPENDIX C: Correspondence from the Army	296
	APPENDIX D: Interaction of Zinc and Cadmium and Toxicity of Zinc Cadmium Sulfide Activators	299

APPENDIX E:	Public Meetings Agendas	**302**
APPENDIX F:	Sampling and Analytic Methods for Zinc Cadmium Sulfide	**311**
APPENDIX G:	Review of AEHA Risk Assessment Reports on Zinc Cadmium Sulfide	**316**
APPENDIX H:	Review of EPA, ATSDR, and CDC Comments on the Army's Risk Assessment Reports on Zinc Cadmium Sulfide	**335**
APPENDIX I:	Cadmium Exposure Assessment, Transport, and Environmental Fate	**338**
GLOSSARY		**361**

Toxicologic Assessment of the Army's Zinc Cadmium Sulfide Dispersion Tests

Summary

BECAUSE OF CONCERN over the possible use of biologic warfare (BW) by a foreign power against the United States and its allies, President Roosevelt in 1942 established the U.S. Biological Warfare Program. During the 1950s and 1960s, Stanford University and other contractors for the U.S. Army Chemical Corps conducted dispersion tests using fluorescent particles of zinc cadmium sulfide (ZnCdS) as part of the BW program in Minneapolis, MN; Corpus Christi, TX; Fort Wayne, IN; St. Louis, MO; and 29 other urban and rural locations in the United States and Canada. The ZnCdS tests were conducted to determine how BW agents disperse in various environments and to determine the munitions requirement (the quantity of a material required to achieve a particular military objective) for the strategic use of BW agents against selected cities and other areas. The ZnCdS particles were not themselves BW agents, but rather were nonbiologic simulants of bacterial particles. The compound was considered to be desirable as a nonbiologic simulant for the following reasons: it fluoresces under ultraviolet (UV) light and therefore can be easily detected; its particle diameter (2-3 µm) and mass, and thus its behavior in air, are similar to those of BW agents; it is economically feasible to use; it was thought to be nontoxic to humans, animals, and plants; and it is relatively stable in the atmosphere.

Upon learning of the dispersion tests in the early 1990s, government officials and citizens in cities where the tests had occurred raised concerns about the thousands of people who might unknowingly have been exposed to ZnCdS. After some information on the tests became public, people living in areas where the tests had been conducted attributed various illnesses, including cancer and reproductive difficulties, to exposure to the chemical.

In response to the initial expressions of concern from the residents and from Congressional representatives of Minneapolis and Corpus Christi and senators from Minnesota and Indiana, the U.S. Army Environmental Hygiene Agency (AEHA) prepared reports that retrospectively assessed the health risk to humans who had been exposed to ZnCdS in those cities. The AEHA assessments were based on a review of the toxicity of cadmium because little information on the toxicity of ZnCdS was available in the scientific literature or in Army files and because AEHA considered cadmium to be the most-toxic component of ZnCdS. AEHA did not believe that zinc, an essential nutrient that is toxic only at high doses, would contribute to the toxicity of ZnCdS. In the AEHA reports, human exposures to cadmium were estimated from ZnCdS-exposure monitoring data recorded at the time of the releases. AEHA concluded: "Conservative evaluation of the available data using EPA risk assessment methodology and comparisons with available standards, ambient air data, and health effects information indicates that the measured concentrations in the test areas should not have been associated with any adverse health effects for residents in the test areas. The estimated excess cancer risks are much less than the risk levels generally considered acceptable by the EPA. These comparisons and evaluations indicate that the ZnCdS tests posed negligible health threats to residents of the test areas."

TASKS OF THE SUBCOMMITTEE

In July 1994, the Army asked the National Research Council to independently review the AEHA reports assessing the health risks for Corpus Christi and Minneapolis, determine the reasonableness of the AEHA conclusions, and, if necessary, suggest recommendations for improving the assessments. In the fiscal year 1995 Department of Defense appropriations, Congress responded to the public-health concerns by directing the secretary of defense to request an independent study by the Research Council concerning the possible adverse health effects of human exposures to ZnCdS as a result of the Army's dispersion tests. Consequently, the Army asked that the Research Council study already under way be substantially expanded to determine independently the health risks associated with exposure to ZnCdS in all exposed U.S. locations, hold public meetings in selected cities where the ZnCdS tests had been conducted, and review the environmental fate of ZnCdS.

The National Research Council assigned the project to the Committee on Toxicology (COT) of the Board on Environmental Studies and Toxicology in the Commission on Life Sciences. COT convened the Subcommittee on Zinc Cadmium Sulfide, which conducted the study and prepared this report. The subcommittee members were chosen because of distinguished expertise in toxicology, medicine, pathology, epidemiology, pharmacology, chemistry, environmental health, environmental fate, industrial hygiene, ecology, biostatistics and mathematical modeling, risk assessment, risk communication, and interpretation of technical information. The subcommittee was charged with the following tasks:

1. Determine the appropriateness of using cadmium-toxicity data as a surrogate for ZnCdS-toxicity data in the AEHA's risk-assessment reports.
2. Assess the transport and environmental fate (but not ecologic effects) of ZnCdS.
3. Assess the adequacy of the AEHA estimates of exposures to ZnCdS in Minneapolis, Corpus Christi, Fort Wayne, and other test areas.
4. Review the toxicokinetics of ZnCdS and its surrogate, cadmium.
5. Assess the toxicity of ZnCdS and cadmium (or cadmium compounds), including effects on sensitive human populations, toxicologic interactions of zinc and cadmium, and the toxicologic implications of the variable composition of ZnCdS.
6. Determine the utility and feasibility of conducting an epidemiologic study of the ZnCdS exposures in question.
7. Review the comments of the U.S. Environmental Protection

Agency, Centers for Disease Control and Prevention, and Agency for Toxic Substances and Disease Registry on AEHA's risk-assessment reports on ZnCdS.

8. Identify research gaps in the available information and develop priorities for research.

The subcommittee has produced three reports: (1) an interim report, published in September 1995, which contains the subcommittee's preliminary toxicity assessment of ZnCdS exposures; (2) a final technical report (the present report), which addresses all tasks listed above; and (3) a separately published nontechnical report for the general public to communicate the extent of risk from exposure to ZnCdS.

It should be noted that although some tests involved simultaneous exposures to ZnCdS and biologic simulants, such as *Serratia marcescens* or *Bacillus globigii*, the subcommittee did not assess the implications of such coexposures, because that was beyond its charge and ability. The subcommittee also did not address whether testing of the chemicals without the knowledge of the American public was ethical or violated the public trust; this question is important but was also beyond the subcommittee's charge and expertise.

SOURCES OF INFORMATION

INPUT FROM THE PUBLIC

The subcommittee held three public meetings as part of its evaluation of the possible adverse health effects of human exposures to ZnCdS. The meetings were held in Minneapolis on May 25, 1995; in Fort Wayne on July 31, 1995; and in Corpus Christi on October 18, 1995. Their purpose was to gather information and learn about public concerns related to the releases of ZnCdS. The three cities were chosen because of expressions of concern about possible health effects and the presence of community groups that were gathering information and seeking answers. In addition, the Army had completed risk assessments for each of the three communities.

Questions and issues raised during the public meetings fell into three general categories: concern about possible health effects of exposures to ZnCdS, outrage about being exposed to a chemical by the government without being informed, and requests for information about the spraying activities—how, how much, when, where, and why.

The information presented in the public meetings ranged from individual and family health histories to detailed maps with markers showing health problems in areas that were sprayed. A common theme throughout the testimony was the frustration that answers were not available as to why people had suffered health problems. Many people indicated that they did not know what caused their health problems; many did not assert that ZnCdS exposure was the cause. But they wanted the subcommittee to have whatever information might help to evaluate the concerns raised.

The types of health effects reported most often differed among the three communities. For example, reproductive problems were more commonly reported at the public meeting in Minneapolis than elsewhere, whereas cancer was mentioned most often in Fort Wayne and Corpus Christi. Degenerative diseases of the central nervous system, such as Parkinson's disease, were reported, as were other degenerative and metabolic disorders, such as atherosclerosis and heart problems, arthritis, diabetes mellitus, diabetes insipidus, and osteoporosis. A number of nonspecific complaints that do not appear to fit into any of these disease categories were noted, such as the development of cysts, high blood pressure, dizzy spells, coughing, swollen glands, infections, joint swelling, weight gain, fatigue, and nosebleeds.

OPEN LITERATURE AND UNPUBLISHED REPORTS

The subcommittee reviewed the available toxicity and exposure data on ZnCdS, cadmium, and cadmium compounds. The ZnCdS data came from reports available in the open literature; reports from Stanford University and other Army contractors for the ZnCdS tests; material-safety data sheets; and the Army. In its review of the data supplied by the Army, the subcommittee became aware that some of the exposure data from the Army's tests on ZnCdS are missing. The Army was asked to supply the missing data, and it informed the subcommittee that it was unable to find those data because the information being sought is 30-40 years old, and the data on the Army's dispersion tests had not been cataloged. However, the subcommittee feels confident in the large amount of data that it did review and does not believe it likely that the missing data would alter its conclusions. The available data provide information on general exposures at various locations but not on exposures of individuals, which are important in epidemiologic studies.

The subcommittee reviewed only unclassified data available on ZnCdS. The Army has assured the subcommittee in writing that all the relevant data on ZnCdS dispersion tests have been declassified and provided to the subcommittee (letter attached in Appendix C). According to the Army, the only information that was not declassified pertains to a large area coverage study, in which there is information regarding the altitude from which a very small quantity of a BW agent could be dropped and contaminate about 500,000 square miles of the country. The Army felt that declassification of this information could affect national security.

CONCLUSIONS

To assess the possible adverse health effects of exposure to ZnCdS from the Army's dispersion tests, the subcommittee reviewed the physical and chemical properties of ZnCdS; the toxicokinetics, bioavailability, and toxicity of ZnCdS; the toxicity of other selected cadmium compounds; and the exposures related to the tests. The subcommittee also assessed the risks associated with exposures to ZnCdS, which included an assessment of the information presented by the public, and evaluated the utility and feasibility of conducting an epidemiologic study of the ZnCdS exposures in question. The conclusions of the subcommittee are presented below.

PHYSICAL AND CHEMICAL PROPERTIES

The ZnCdS used in the Army studies was composed of about 80% zinc sulfide (ZnS) and 20% cadmium sulfide (CDs). The concentration of copper or silver was about 0.005%. ZnCdS is not just a physical mixture of the two compounds; its constituents—ZnS and CDs—are sintered by heating a mixture of them to about 900°C so that a crystalline lattice structure containing zinc, cadmium, and sulfur is formed. The sintered compound reportedly does not contain pure ZnS or CDs, because the sintering process is highly efficient. It is stable in atmospheric conditions, insoluble in water and lipids, and poorly soluble in strong acids.

TOXICOKINETICS AND BIOAVAILABILITY OF ZNCDS: AVAILABILITY OF CADMIUM FROM ZNCDS

No studies on the toxicokinetics of ZnCdS were found. Because it is poorly soluble in strong acids and insoluble in water and lipids, ZnCdS probably is not absorbed through the skin or gastrointestinal tract. Its lack of solubility also suggests that it is highly unlikely that free cadmium ions would become bioavailable to target organs as a result of inhalation of ZnCdS. However, information is not available on whether ZnCdS might break down in the respiratory tract into more-soluble components, which could be absorbed into the blood.

TOXICITY

The toxicity database on ZnCdS is sparse. No human studies on ZnCdS are available. Animal data indicate that ZnCdS is not acutely toxic when given orally; that finding is consistent with the insolubility of the compound and its apparent lack of bioavailability. ZnCdS was also not found to be a skin or eye irritant in rabbits.

No reports on the toxicity of inhaled ZnCdS are available in the literature. Because the ZnCdS particles used in the Army's dispersion studies were so small, the particles could probably be inhaled and deposited in the deep lung, but no information is available on the potential toxicity of the particles in the lung. It is also not known whether ZnCdS can be broken down by pulmonary macrophages into more-soluble forms of cadmium.

TOXICITY OF SELECTED CADMIUM COMPOUNDS

Faced with the task of evaluating the potential toxicity of ZnCdS, a compound with largely unknown toxic potential but reasonably well-known physical and chemical properties, the subcommittee considered it prudent to examine toxicity and related data on the most toxic component in ZnCdS, cadmium. The toxic potency of cadmium compounds depends on their in vivo solubility and bioavailability. ZnCdS is neither water-soluble nor apparently bioavailable, so the subcommittee believes that the use of toxicity data on soluble cadmium compounds to estimate the toxicity of ZnCdS constitutes a worst-case scenario. In other words, this approach would lead to an overestimate of the risk associated with ZnCdS.

As a general rule, highly in vivo soluble cadmium compounds—such as cadmium chloride ($CdCl_2$), cadmium sulfate ($CdSO_4$), and cadmium oxide (CdO)—are more toxic than the poorly soluble compounds, such as cadmium sulfide (CDs).

Soluble cadmium compounds can be absorbed from the skin, intestinal tract, or respiratory tract into the bloodstream and can be transported throughout the body with the potential for causing systemic injury. With high short-term exposures, only lung toxicity is seen. With chronic exposures, the most-sensitive sites of injury are the kidneys and bones.

The subcommittee believes that the toxicity of ZnCdS is more like that of cadmium sulfide than like that of other cadmium compounds because the crystalline structures of the two compounds are similar, both compounds are insoluble in vivo, and neither compound is bioavailable. The subcommittee, therefore, chose to base its assessment of the potential toxicity of ZnCdS for noncancer health effects on the toxicity of CDs.

Several studies conducted in experimental animals show that the toxicity of CDs is much lower than the toxicity of soluble cadmium compounds. It takes considerably higher CDs concentrations to produce lung toxicity, and substantially less cadmium becomes bioavailable from inhaled or ingested CDs than from the more-soluble cadmium compounds. CDs administered to rats intratracheally produces a minimal inflammatory response in the lung. The estimated no-observed-adverse-effect level for humans exposed to CDs for 2-3 h by inhalation is 594 µg (that is, 513 µg of cadmium). The highest estimated cadmium doses to any individual at the test sites were below this level. The maximal potential cadmium doses in a populated area were: St. Louis, 24.4 µg (in 31 months); Winnipeg, 14.5 µg (in 22 days); Minneapolis, 6.8 µg (in 1 month). Other test locations had lower values. Toxicokinetic studies showed that inhaled CDs is not absorbed from the lungs into systemic circulation. Some 75% of inhaled CDs is exhaled immediately, and about 90% of the remainder is removed slowly from the lungs into the gastrointestinal tract by normal pulmonary clearance processes.

Several epidemiologic studies have shown that occupational exposure to high concentrations of cadmium and cadmium compounds for many years is correlated with lung cancer. In animal studies, all cadmium compounds examined have been found to produce respiratory tract tumors after chronic exposures. The subcommittee reached the following conclusions on the carcinogenicity of cadmium compounds as they relate to the ZnCdS exposures:

- Inhaled cadmium has been shown in occupational studies and laboratory studies of animals to cause lung cancer, but not cancer at other body sites.
- Cadmium inhalation exposures associated with increased lung-cancer risk in animal studies involved higher concentrations (100-1,000 times higher), longer periods (lifetime exposures), and more-soluble compounds than the exposures to cadmium from ZnCdS in the Army's testing program.
- The cancer-potency data available on cadmium are based on relatively high occupational exposures to cadmium compounds of undefined solubility.
- A quantitative risk assessment for lung cancer, based on occupational exposures of humans to cadmium compounds, is likely to overestimate the risk of lung cancer for ZnCdS exposures from the Army's dispersion tests.

EXPOSURE ASSESSMENT

The subcommittee estimated the magnitude of potential human doses of cadmium as a result of the dispersions of ZnCdS by the Army in U.S. and Canadian locations. On the basis of the data available, the maximum estimated cadmium doses (calculated from ZnCdS) in populated areas were about 6.8 µg in Minneapolis, MN; 14.5 µg in Winnipeg, Canada; and 24.4 µg in St. Louis, MO. The maximal potential cadmium inhalation dose to any individual of 24.4 µg recorded in St. Louis is equivalent to living 1-8 months in a typical U.S. city where the cadmium intake is believed to be about 0.1-0.8 µg/d. The daily cadmium intake via inhalation in rural areas is less than 0.02 µg.

It is important to consider doses such as these in the context of typical ambient background exposures. Cadmium is a natural component of the earth's crust. All soils and rocks, including coal, have some cadmium in them. Cadmium is naturally found as a component of small particles present in air. Cadmium enters the air from the burning of coal and household waste, and from metal mining and refining processes. Food, water, and smoking are the largest potential sources of cadmium exposures for the general population. Exposures to normal ambient "background" concentrations of cadmium result in a total daily human cadmium intake of 12-84 µg for an adult. Air contributes 0.02-0.8 µg of cadmium per day, water 2-20 µg/d, food 10-60 µg/d, and smoking 2-4 µg/d. Thus, the subcommittee estimated that the average total daily intake from typical environmental and industrial exposures to cadmium from all media (soil, water, food, and air) in urban areas is greater than the total exposures to cadmium resulting from the Army's ZnCdS tests at most sites. The highest estimated cadmium intake (from inhalation) from the Army's ZnCdS dispersion tests was 24.4 µg in St. Louis. At about half the test sites, the maximal concentrations of airborne cadmium (in the form of ZnCdS) were above the estimated urban average daily airborne cadmium, but the subcommittee believes that these short-term high concentrations would have had minimal impact on total cadmium exposure, which is mainly from water, food, and soil.

RISK ASSESSMENT OF ZNCDS EXPOSURES

The subcommittee concluded that only the respiratory tract was potentially at risk from exposure to particles of ZnCdS because the particles of ZnCdS were respirable and would be expected to deposit deep in the lungs.

However, results of several toxicity studies indicate that CDs (the cadmium compound considered to be most relevant for assessing noncancer ZnCdS toxicity) has little toxicity in the respiratory tract. Rats exposed to CDs at 39,600 mg-min/m^3 (equivalent to total inhaled dose of 594 mg of CDs or 513 mg of cadmium) had only a mild pulmonary response. By using an uncertainty factor of 1,000 (513 mg/1,000 = 513 µg), one would not expect adverse health effects, even in sensitive populations, from exposure to the 513-µg-dose level. The highest estimated cadmium doses at the test sites were below 513 µg of cadmium. The subcommittee concludes that the exposures to ZnCdS from the Army's tests (for which data are available) should not have caused noncancer health effects in exposed persons.

The subcommittee also estimated maximal lifetime excess lung-cancer risk for the test locations. Test sites with the highest cadmium exposures were Biltmore Beach, St. Louis, Winnipeg, and Minneapolis. Cancer risk estimates were based on the cadmium content of ZnCdS because no studies on the carcinogenicity of ZnCdS have been conducted, and cadmium is the most-toxic component, so this approach is likely to overestimate the risk. For each city, the location with the highest average reported air concentration of cadmium was used to calculate cancer risk. The risk estimates were multiplied by a factor of 10 to account for exposure to sensitive populations. The estimated upper-bound lung-cancer risks ranged from less than 0.01×10^{-6} to 24.0×10^{-6} (0.4 to 24 per million).

The vast majority of people in the test areas were exposed to concentrations that were less than the highest by a factor of 2-10 or more. Accordingly, their cancer risks would be lower by a factor of 2-10 or more. The subcommittee concluded that given the extremely low concentrations of ZnCdS (or cadmium from ZnCdS) in the air and the short duration of exposure, the lung-cancer risk, if any, is likely to be very low. Thus, it is unlikely that anyone in the test areas would have developed lung cancer from direct exposure to airborne releases of ZnCdS.

On the basis of a review of the scientific literature, information presented at the public meetings, and an analysis of exposure information, the subcommittee concluded that the list of disorders reported by people living in areas where ZnCdS was released reflects diseases found in the general population and cannot be attributed to the ZnCdS releases.

SCIENTIFIC FEASIBILITY OF EPIDEMIOLOGIC STUDY

The subcommittee concluded that an epidemiologic study of the affected populations is not feasible. There are three major barriers to carrying out an epidemiologic study of the health effects of ZnCdS: lack of complete and accurate exposure data on individuals; inadequacies in data on health outcomes before, during, and after the periods of exposure; and, because of the low exposures, the requirement of a very large sample to detect any small increase in adverse health effects. Accurate measurement of individuals' doses is not possible, given the very small contribution of ZnCdS to the concentration of cadmium in the environment. Information on potential confounders also would be lacking.

RECOMMENDATIONS

It is highly unlikely that free cadmium ions would become bioavailable to target organs as a direct result of inhalation of ZnCdS. However, information is not available on whether ZnCdS might break down in the respiratory tract into more-soluble components, which could be absorbed into the blood.

- The subcommittee recommends that the Army conduct studies to determine the bioavailability and inhalation toxicity of ZnCdS in experimental animals. This research will strengthen the database needed for risk assessment of ZnCdS and lessen the need to rely on the use of cadmium or cadmium compounds as surrogates for toxicity information.
- The subcommittee recommends that when the results of the research become available, they be reviewed by experts outside the Army to determine whether the subcommittee's conclusions are still valid or should be modified.

People were outraged at being exposed to chemicals by the government without their knowledge or consent.

- The subcommittee did not address ethical and other social issues about the Army's dispersion tests; these questions are important, and the Army must develop a mechanism for addressing the public's sense of outrage, but these issues were beyond the subcommittee's charge and expertise.

Toxicologic Assessment of the Army's Zinc Cadmium Sulfide Dispersion Tests

I

INTRODUCTION

BECAUSE OF WIDESPREAD CONCERN over possible use of biologic warfare (BW) by a foreign power against the United States and its allies, President Roosevelt in 1942 established the U.S. Biological Warfare Program. The policy of the United States concerning BW between 1941 and 1973 was, first, to deter its use against the United States and its allies and, second, to retaliate by using pathogenic agents if deterrence failed (U.S. Army 1977). The stockpile of offensive BW agents was destroyed in 1973 in accordance with a directive issued by President Nixon in 1969.

The potential tactical use of BW agents required the development of tables of munitions requirements (the quantities of materials required to achieve particular military objectives) for the strategic use of BW agents against specific cities. To obtain some estimate of the amount of BW material required to meet certain objectives in given cities, an indirect approach was used: to simulate the BW agents, and run tests with the simulants in suitable locations. The U.S. Army used both biologic and nonbiologic simulants in its tests. Biologic simulants are defined as living microorganisms that are not normally capable of causing infection, that represent the physical and biologic characteristics of potential pathogenic microbiologic agents, and that were considered medically safe for operating personnel and the general public. Nonbiologic simulants are nonliving inert (usually inorganic) materials; they are formed to resemble the size of BW agents for dispersion in the air in a manner similar to BW agents, but they are not themselves BW agents.

The biologic simulants used included *Serratia marcescens, Bacillus globigii, Bacillus subtilis,* and *Aspergillus fumigatus*; all these were considered by the Army to be safe at the time of their use (U.S. Army 1977). (More recent research indicates that the organisms are generally nonpathogenic or of low virulence in normal populations, but some could become pathogenic in immunocompromised persons.) The nonbiologic simulants used included zinc cadmium sulfide (ZnCdS) and sulfur dioxide (SO_2), the former of which is the subject of this report. Although some of the Army's tests involved exposures to ZnCdS at the same time as biologic agents, such as *Serratia marcescens*, the subcommittee did not assess the implications of such coexposures, because that was beyond its charge and ability. Information on interactions between biologics and chemicals can be found in the Institute of Medicine's 1996 report entitled "Interactions of Drugs, Biologics, and Chemicals in the U.S. Military Forces." Appendix A provides greater detail on the U.S. BW program and the selection of biologic and nonbiologic simulants.

ZnCdS DISPERSION TESTS

During the 1950s and 1960s, Stanford University and other contractors for the U.S. Army Chemical Corps conducted dispersion tests with fluorescent particles of ZnCdS as part of the BW program in Minneapolis, MN; Corpus Christi, TX; St. Louis, MO; Fort Wayne, IN; and 29 other urban and rural locations in the United States and Canada. Tables 5-1 and B-1 show the locations, dates, and amounts of ZnCdS released. Appendix A provides greater detail on and the reasons for the selection of the test cities or other test locations.

INTRODUCTION

ZnCdS is a sintered chemical (it is formed by heating a mixture of zinc sulfide, ZnS, and cadmium sulfide, CdS) composed of 80% ZnS and 20% CdS. It was considered to be desirable as a nonbiologic simulant for several reasons: (1) it fluoresces under ultraviolet (UV) light and therefore can be easily detected, (2) its particle size (0.5-6.25 µm), which is similar to that of several microorganisms, makes it easily dispersible (Leighton 1955), (3) it was economically feasible, (4) it was perceived to be nontoxic to humans, animals, and plants, and (5) it is stable in the atmosphere.

The ZnCdS tests were initially conducted in Minneapolis, St. Louis, and Winnipeg. These cities were selected because of their similarity—in meteorologic, terrain, population, and physical characteristics—to such cities in the former Soviet Union as Moscow and St. Petersburg (formerly Leningrad). As the field testing program progressed, it was recognized that data were needed for other geographical areas. Tests with single and multiple munitions were conducted in forests—in mountainous areas of Targhee National Forest in Idaho; in moderately to heavily covered flat land in Florida, and tropical islands off the Panama coast. In addition, numerous field tests were conducted by the British in India and Australia which added to the variety of terrain and geographical variables. When all of these tests were completed, there was an abundance of information showing directly how various biologic agents behaved under a variety of conditions which were either representative of or bracketed the conditions experienced in areas of combat.

Operation LAC (for Large Area Coverage) was the largest test program ever undertaken by the Army's Chemical Corps. The test area covered the United States from the Rockies to the Atlantic, from Canada to the Gulf of Mexico. The tests proved the feasibility of covering large areas (thousands of square miles) of a country with BW agents based on ZnCdS particles as simulants.

PUBLIC CONCERN IN RESPONSE TO ZnCdS DISPERSION TESTS

On learning of the dispersion tests in the early 1990s, government officials and citizens in several places, such as Minneapolis, MN, Corpus Christi,

TX, and Fort Wayne, IN, raised concerns about the thousands of people who might unknowingly have been exposed to ZnCdS. After information on the tests became public, some people living in areas where the tests were conducted attributed various illnesses, including cancer and reproductive difficulties, to exposure to the chemical.

In response to the public concern, the AEHA prepared reports that assessed the health risk to humans exposed to ZnCdS in Minneapolis, Corpus Christi, Fort Wayne, and St. Louis (AEHA 1994, 1995a,b). The assessments were based on a review of the toxicity of cadmium because little information on the toxicity of ZnCdS was available in the scientific literature or in Army files and because AEHA considered cadmium to be the most toxic component of ZnCdS. Cadmium, a toxic metal, is present in the ambient environment (in food, air, and water). Zinc is an essential nutrient (NRC 1980) and is toxic only at high doses; AEHA did not believe that zinc would contribute to any toxicity that ZnCdS might have. In the AEHA reports, human exposure to cadmium was estimated from the ZnCdS-exposure monitoring data recorded at the time of the releases. AEHA concluded: "Conservative evaluation of the available data using EPA risk assessment methodology and comparisons with available standards, ambient air data, and health effects information indicates that the measured concentrations in the test areas should not have been associated with any adverse health effects for residents in the test areas. The estimated excess cancer risks are much less than the risk levels generally considered acceptable by the EPA. These comparisons and evaluations indicate that the ZnCdS tests posed negligible health threats to residents of the test areas" (AEHA 1994, 1995a,b).

TASKS OF THE SUBCOMMITTEE

In July 1994, the Army asked the National Research Council to review the AEHA reports assessing the health risks for Corpus Christi, TX, and Minneapolis, MN (AEHA 1994), determine the reasonableness of the conclusions, and, if necessary, suggest recommendations for improving the assessments. In the fiscal year 1995 Department of Defense appropriations, Congress responded to growing public-health concerns by directing the secretary of defense to request an in-depth independent study by the Research Council of the possible adverse health effects of human exposure to ZnCdS as a result of the Army's dispersion tests. Consequently, the Army asked that the Research Council study on ZnCdS already under way be expanded substantially to determine the health risks associated with exposure to ZnCdS in all exposed U.S. locations, hold public meetings in selected cities where ZnCdS tests were conducted, and review the environmental fate of ZnCdS.

The Research Council assigned the project to the Committee on Toxicology (COT) of the Board on Environmental Studies and Toxicology in the Commission on Life Sciences. COT convened the Subcommittee on Zinc Cadmium Sulfide, which prepared this report. The subcommittee members were chosen for their recognized expertise in toxicology, medicine, pathology, epidemiology, pharmacology, chemistry, environmental health, environmental fate, industrial hygiene, ecology, biostatistics and mathematical modeling, risk assessment, risk communication, and interpretation of technical information. The subcommittee was charged with the following tasks:

1. Determine the appropriateness of using cadmium-toxicity data as a surrogate for ZnCdS-toxicity data in the AEHA's risk-assessment reports.
2. Assess the transport and environmental fate (but not ecologic effects) of ZnCdS.
3. Assess the adequacy of the AEHA estimates of exposures to ZnCdS in Minneapolis, Corpus Christi, Fort Wayne, and other test areas.
4. Review the toxicokinetics of ZnCdS and its surrogate, cadmium.
5. Assess the toxicity of ZnCdS and cadmium (or cadmium compounds), including effects on sensitive human populations, toxicologic interactions of zinc and cadmium, and the toxicologic implications of the variable composition of ZnCdS.
6. Determine the utility and feasibility of conducting an epidemiologic study of the ZnCdS exposures in question.
7. Review the comments of the U.S. Environmental Protection Agency, Centers for Disease Control and Prevention, and Agency for Toxic Substances and Disease Registry on AEHA's risk-assessment reports on ZnCdS.

8. Identify research gaps in the available information and develop priorities for research.

The subcommittee's task was to produce three reports: (1) an interim report, which was published in September 1995 (NRC 1995) and contains the subcommittee's preliminary health risk assessment of ZnCdS exposures, (2) a comprehensive final technical report (the present report), which addresses all the 8 issues listed above, and (3) a nontechnical summary report to disseminate the findings to the general public, which is published separately (NRC 1997).

Although some tests involved simultaneous exposures to ZnCdS and biologic simulants, such as *Serratia marcescens* or *Bacillus globigii*, the subcommittee did not assess the implications of such coexposures, because that was beyond its charge and expertise. The subcommittee also did not address whether testing of the simulants without the knowledge of the American public was ethical or violated the public trust; this question is also important but is beyond the subcommittee's charge and expertise.

SOURCES OF DATA

The subcommittee held public meetings to gather information in selected cities in which dispersion tests were conducted. Three public meetings were held—in Minneapolis, Fort Wayne, and Corpus Christi. The public was informed about the meetings through advertisements in local newspapers and through the courtesy of local radio and television stations. A total of several hundred persons attended the meetings, and they were covered by television, radio, and newspapers. Information gathered and issues and concerns raised at the public meetings are presented and addressed in Chapter 2, which contains the details of the three public meetings.

The subcommittee reviewed toxicity data on ZnCdS, cadmium, cadmium compounds, and zinc. The ZnCdS data came from reports available in the open literature; reports from the Army's contractors for the ZnCdS tests; material-safety data sheets; and the U.S. Army. The subcommittee is aware that some of the exposure data from the Army's tests on ZnCdS are missing. The Army was asked to supply the missing data, and it has informed the subcommittee that it is unable to find the missing data because the information being sought is 30-40 years old, and the data on the Army's dispersion tests were not cataloged. Although some exposure data from the Army's ZnCdS dispersion tests are missing, the subcommittee feels confident in the quality of the large amount of data that it reviewed and does not believe it likely that the additional missing data would alter its conclusions. These data provide information on exposures at various locations but not on exposures to individuals, which is important in epidemiologic studies.

The subcommittee reviewed only the unclassified data available on ZnCdS. The Army has assured the subcommittee in writing that all the relevant data on ZnCdS dispersion tests have been declassified and provided to the subcommittee (letter attached in Appendix C). According to the Army, the only information that was not declassified pertains to the LAC study, on which there is information regarding the altitude from which a very small quantity of a BW agent could be dropped and contaminate approximately 500,000 square miles of the country. The Army felt that declassification of this information could affect the national security of this country (Col. Robert Gum, U.S. Army Medical Research Institute of Chemical Defense, personal commun., 1995).

A vast amount of data on cadmium toxicity is available in the open literature; the subcommittee drew heavily on the detailed toxicity reviews conducted by the Agency for Toxic Substances and Disease Registry (ATSDR 1993), International Agency for Research on Cancer (IARC 1993), World Health Organization (IPCS 1992), and Occupational Safety and Health Administration (OSHA 1992).

The subcommittee reviewed the toxicity data on zinc and concluded that it would not be toxic at the low exposure levels associated with ZnCdS dispersion tests. The subcommittee also reviewed the possible interaction of zinc and cadmium (Appendix D). The toxicities of minor components of ZnCdS, such as copper ($\approx.005\%$) and silver ($\approx.005\%$) were also reviewed (Appendix D).

STRUCTURE OF THE REPORT

The subcommittee assesses the health risk associated with ZnCdS exposures in various U.S. locations by conducting an in-depth review of the toxicity and toxicokinetic data on ZnCdS and its most toxic component—cadmium—by estimating the maximal concentration of cadmium in each test location and by characterizing the risk associated with the estimated maximal cadmium exposures. The subcommittee considered total exposures to cadmium as a consequence of exposure to ZnCdS from all routes—inhalation and ingestion of food and water—in relation to known background exposures to cadmium. The effect of ZnCdS and cadmium on potentially susceptible groups in the human population was also considered. In addition, the subcommittee assessed the environmental fate of ZnCdS and determined the utility and feasibility of conducting an epidemiologic study of populations exposed to it.

Chapter 2 presents the information gathered from the public meetings. Chapter 3 reviews the toxicity and related data on ZnCdS, and Chapter 4 evaluates the toxicity, environmental fate, and epidemiology data of cadmium compounds. Chapter 5 evaluates the exposures to ZnCdS and cadmium, and Chapter 6 contains the subcommittee's risk assessments of ZnCdS for noncancer and cancer effects. Chapter 7 discusses the feasibility of conducting epidemiologic studies of ZnCdS-exposed persons. Chapter 8 contains the subcommittee's conclusions and recommendations. Appendix A reviews the historical background of the U.S. Biological Warfare Program. Appendix B summarizes the doses and concentrations of ZnCdS particles from the Army's dispersion tests. Appendix C contains correspondence with the U.S. Army. Appendix D discusses the interaction of zinc and cadmium and the toxicity of copper and silver. Appendix E provides information concerning public meetings. Appendix F reviews sampling and analytic methods for ZnCdS. Appendix G contains a review of the Army Environmental Hygiene Agency's risk-assessment reports on ZnCdS, and Appendix H reviews the comments of EPA, ATSDR, and CDC on the Army's risk-assessment reports. Appendix I discusses the exposure assessment for cadmium. The final section of the report—the glossary—provides brief descriptions of technical terms used in the report.

2

INPUT FROM THE PUBLIC

THE SUBCOMMITTEE HELD 3 PUBLIC MEETINGS as part of its evaluation of the health impacts from the release of zinc cadmium sulfide (ZnCdS) by the Army. The meetings were held in Minneapolis, MN, on May 25, 1995; Fort Wayne, IN, on July 31, 1995; and Corpus Christi, TX, on October 18, 1995. Their purpose was to gather information and learn about public concerns related to the release of ZnCdS. Minneapolis and Fort Wayne were chosen because of heightened concern about possible health effects and the presence of community groups that were gathering information and pressing for answers. Corpus Christi was chosen because it was a coastal city and the subcommittee wanted to determine whether anything more could be learned from the site of an aerial coastal spray. In addition, the Army had completed risk assessments for each of the 3 communities.

Each meeting was held at a location that was easily accessible to the public. Before the meeting, National Research Council staff sent press releases to all the local media and to national media that cover Research Council activities in the area and advertised the meeting in local newspapers. Mayors, members of Congress representing the district, U.S. senators of the state, and other elected officers were informed of the meetings so that they could inform their constituents. People were encouraged to register to speak by calling or writing to Kulbir Bakshi, project director for the subcommittee. The subcommittee was available before and after the meeting to talk to the media.

Each meeting began with a presentation about the subcommittee, which included background on the Research Council and the National Academy of Sciences, how the subcommittee was formed, the tasks assigned to it, a brief description of the work already completed by it, and the introduction of its members.

The public meetings in Minneapolis and Fort Wayne each lasted a full day. Of the 88 people who signed the registration sheets in Minneapolis, 24 presented formal comments. Of the 162 who signed the registration sheets in Fort Wayne, 33 spoke. The meeting in Corpus Christi lasted a half-day; of the 23 who signed up, 11 spoke. Appendix E contains the agendas for the public meetings. After each presentation, the subcommittee members were able to ask questions to clarify comments and gain additional information. At the conclusion of the comment sessions, open-microphone sessions provided an opportunity for an informal exchange of information and points of view among participants, members of the audience, and subcommittee members. The proceedings of the meetings were recorded and transcribed. Fact sheets provided at the meetings included information about how to obtain transcripts.

NATURE OF COMMENTS

The public meetings were both emotional and informative. The many questions and issues raised were in three general categories: concern about possible health effects of exposure to ZnCdS, outrage about being exposed to a chemical by the government without being informed, and requests for information about the spraying activities—how, how much, when, where, and why.

Many people expressed pride in being American and having served their country but difficulty in believing that their government could have carried out the spraying activities without disclosing them to the public. People expressed concern about identifying and holding accountable those responsible for the testing and about ensuring that testing without the informed consent of the subjects would never be repeated. Those concerns are related to trust in the government. Several people who presented comments clearly stated that an open, thorough examination of the issues, beginning with the work of this subcommittee and its open-meeting process, could help to restore confidence in the government.

It was evident that many speakers in the three communities had put a great deal of effort into preparing their comments. Some had compiled and documented detailed information about the health status of those who were in the area during the release of ZnCdS or had researched other aspects of community health. The information presented ranged from discussion of individual or family health histories to detailed maps with markers showing health problems in areas that were sprayed.

In Fort Wayne, one family described how the mother had developed malignant melanoma and had been sent to Minneapolis for special treatment. Since becoming aware of the ZnCdS spraying, they had discovered that between 1973 and 1976, 17 other people from Fort Wayne had been sent to the same clinic in Minneapolis for treatment of melanoma. One of the sisters in the family had put together a map showing the homes of 70 persons who either had died from cancer or had had cancer diagnosed, the location of ZnCdS sampling stations, and the general pattern of the fallout of dispersed ZnCdS.

Several speakers presented lists of people with health problems who lived in areas where ZnCdS had been released. In Corpus Christi, for example, a woman presented a list of 80 people, all with cancer, who lived in the town of Taft, 18 miles north of Corpus Christi, 90% of them lived in a 12-block area of the town. The cancers in the town included cancers of the liver, brain, breast, colon, and prostate. The speaker was hoping for insight into what could have caused this group of cancers.

Some of the individual comments were very personal and emotional. For example, a mother in Minneapolis who lived in the area where ZnCdS had been released discussed the physical problems of her daughter who died at the age of 46 after a long history of health problems, including irregular menstrual cycles, repeated growth of very large cysts, and a collapsed lung.

Many people indicated that they did not know what caused their health problem, nor did they assert that ZnCdS exposure was the cause, but they wanted the subcommittee to have whatever information might help to answer the concerns raised.

A common theme throughout the testimony was the frustration that answers were not available as to why people had their health problems. People described having rare illnesses and seeing several experts to look for causes of their illnesses. Many wondered whether ZnCdS, which they understood to be carcinogenic and had been released in an area where they lived, was in some way responsible for their health condition.

Many questions were raised about individual health problems at the public meetings and in written comments submitted to the subcommittee. In general, the questions were divided into four categories: about whether exposure to ZnCdS could have caused the problems, about the unique sensitivity of children and whether children would be more sensitive to exposure to ZnCdS than adults, about future health problems both for the participants and for their children and their children's children, and about cumulative health effects. The subcommittee's report for the general public addresses the most commonly asked questions (NRC 1997).

The comments and testimony at the public meetings made clear a serious need for public-health information in the communities that was not being addressed. People had many questions but did not know where to go for health-care information. People reported that inquiries to local, state, and federal health agencies had provided little information.

Many of the comments about possible effects on human health and the environment have been useful in the subcommittee's analysis. For example, the community concerns about a wide array of health effects showed the subcommittee that it had to consider many possible health effects and not focus only on cancer. Comments also emphasized that the subcommittee's analysis should consider particularly sensitive (susceptible) populations (such as children and the elderly), variations in populations, and age distribution. Descriptions of conditions in the communities at the time of testing increased the subcommittee's insight into possible community-specific exposure pathways, such as the use of surface-water reservoirs to supply drinking water, and variations in life styles. Some, especially representatives of the Leech Lake Band of Chippewa in Minnesota, emphasized the need to consider effects on the environment, not just those on human health. However, the effect of ZnCdS on the environment is beyond the scope of the subcommittee's task.

Some of the speakers remembered the testing, and a few had helped to carry it out. They were able to provide insight into how the material was dispersed and how the sampling was conducted and they provided their recollections of what happened in the communities during the tests. Some people who were involved in the conduct of the ZnCdS dispersion tests did not report adverse health effects associated with the tests.

SPECIFIC HEALTH PROBLEMS

The types of health effects discussed most often by concerned residents differed among the three communities. For example, reproductive problems were more commonly reported at the public meeting in Minneapolis than elsewhere, whereas cancer was mentioned most often in Fort Wayne and Corpus Christi.

Overall, the most common reports presented at the public meetings were on cancer. All three communities reported concern about apparent cancer clusters, although the type of cancer varied. In Minneapolis, for example, an organized community group called Children of the Fifties reported cancer and other health problems in children who attended the Clinton Elementary School. ZnCdS had been released from the roof of the school and from nearby mobile vehicles. The group had identified 350 of the 800 students who attended the school and found that many had already died, a large number from cancer. The group is considered in more detail in Chapter 7. In Fort Wayne, residents described a number of cases of malignant melanoma and breast cancer. In particular, some specific cases of breast cancer in young women were noted by citizens and a local physician. The cancers reported at the public meetings included cancers of the prostate, thyroid, salivary glands, esophagus, breast, bladder, colon, kidney, uterus, ovaries, blood cells, and brain. Other health problems reported in the three communities included difficulty in getting pregnant; miscarriages and other reproductive disorders, including enlarged uterus and female infertility; respiratory problems, including pneumonia, bronchitis, and asthma; thyroid disorders; swollen glands;

joint pain; and skin problems. Other disease categories identified were congenital disorders, including mental retardation, learning disorders, Down's syndrome, and liver malformations. Immunologic diseases included disorders of the thyroid gland and skin and asthma. Infections of lungs and ears were also reported.

Degenerative diseases of the central nervous system, such as Parkinson's disease, were reported, as were other degenerative and metabolic disorders such as atherosclerosis and heart problems, arthritis, diabetes mellitus, diabetes insipidus, and osteoporosis.

A number of nonspecific complaints that do not fit into any disease category were noted, such as the development of cysts, high blood pressure, dizzy spells, coughing, swollen glands, joint swelling, weight gain, fatigue, and nosebleeds.

CONCLUSIONS

The subcommittee has drawn valuable information and guidance from the presentations made at the public meetings and other comments submitted by the public. Members of the public who presented comments to the subcommittee were concerned about a wide array of health effects, including cancer and reproductive effects, and about whether these effects could have been caused by exposure to ZnCdS. They were also concerned about the possibility of increased risk to sensitive populations, such as children and the elderly. Moreover, people were outraged at being exposed to chemicals by the government without their knowledge or consent. The subcommittee did not address ethical and other social issues about the Army's dispersion tests; these questions are important, and the Army must develop a mechanism for addressing the public's sense of outrage, but these issues were beyond the subcommittee's charge and expertise.

In the remainder of this report, the subcommittee describes its attempt to determine whether diseases reported by people living in areas where ZnCdS was released are caused by ZnCdS. The subcommittee determined the biological plausibility of the diseases reported by the public to be caused by the exposures to ZnCdS and determined the extent to which the exposures might have caused an increase in the diseases above background levels.

3

TOXICITY AND RELATED DATA ON ZINC CADMIUM SULFIDE

PHYSICAL AND CHEMICAL PROPERTIES

ZINC CADMIUM SULFIDE (ZnCdS) (CAS NO. 68583-45-9) was manufactured originally by the New Jersey Zinc Company and later by the U.S. Radium Corporation in New Jersey. The ZnCdS used in the Army studies appeared as a yellow or green (depending on whether copper or silver was used as an activator), somewhat fluffy powder under visible light. It was composed of about 80% zinc sulfide, ZnS, and 20% cadmium sulfide, CdS; the concentration of copper or silver in ZnCdS was about 0.005%. Magnesium silicate or some other silicate was added at 1-2% to facilitate the dispersion of ZnCdS. ZnCdS is not just a physical mixture of the two compounds; its constituents—ZnS and CdS—are sintered by heating a mixture of the 2 compounds to about 900°C (Ruda 1992) so that a crystalline lattice structure containing zinc, cadmium, and sulfur is formed. The sintered compound reportedly does not contain pure ZnS or CdS, because the sintering process is highly efficient (Sheila Fabiano, USR Optonix, Inc., personal commun., May 27, 1995).

Zinc sulfide exists in two crystalline forms: a cubic form with the zincblende structure (the most common form in nature, occurring as the rather common mineral sphalerite) and a hexagonal form, which is sometimes referred to as the wurtzite structure (Fedorov and others 1993). The cubic form forms at lower temperatures than the hexagonal. The hexagonal form is stable at formation temperatures above 1026°C. Both forms have tetrahedral coordination about both the zinc and sulfur atoms, with each sulfur atom tetrahedrally bonded to four zinc atoms and each zinc atom tetrahedrally bonded to four sulfur atoms. Figure 3-1 shows diagrams of both forms. Cadmium sulfide exists primarily in the hexagonal (wurtzite) form and occurs in nature as the very rare mineral greenockite. Solid solutions of ZnS and CdS can exist in either crystal form, depending on the composition, sintering temperature, and cooling rate. At the composition used in the Army's studies (80% ZnS and 20% CdS), the transition between wurtzite and zincblende occurs at about 500°C. However, rapid quenching of the material from the reported 900°C sintering temperature would preserve the wurtzite form.

ZnCdS is a fluorescent material and fluoresces brilliantly under ultraviolet light chiefly in the region from 3100 Å to 4000 Å.

The density of ZnCdS is 4.0 g/cm^3 (Leighton and others 1965). A gram of ZnCdS contains some 10^{10} particles. It can be made in the form of particles small enough to meet the criteria for studying atmospheric diffusion, and the dispersibility of ZnCdS is high enough to enable its practical use in the Army's dispersion tests (Leighton 1955). ZnCdS used in the Army's tests had a mass median aerodynamic diameter of 2-3 µm, and the particle diameter ranged from 0.5 to 6.25 µm.

ZnCdS is stable in the atmosphere long enough to conduct the tracer experiment (hours to days). It is insoluble in water and only weakly soluble in strong acids, but it is soluble in oxidizing acids, such as concentrated nitric acid. ZnCdS continues to fluoresce after 2 h of exposure to air at 450°C. When immersed in 12 M hydrochloric acid, it is not completely destroyed until after some 8 h (Leighton 1955). ZnCdS is insoluble in lipids.

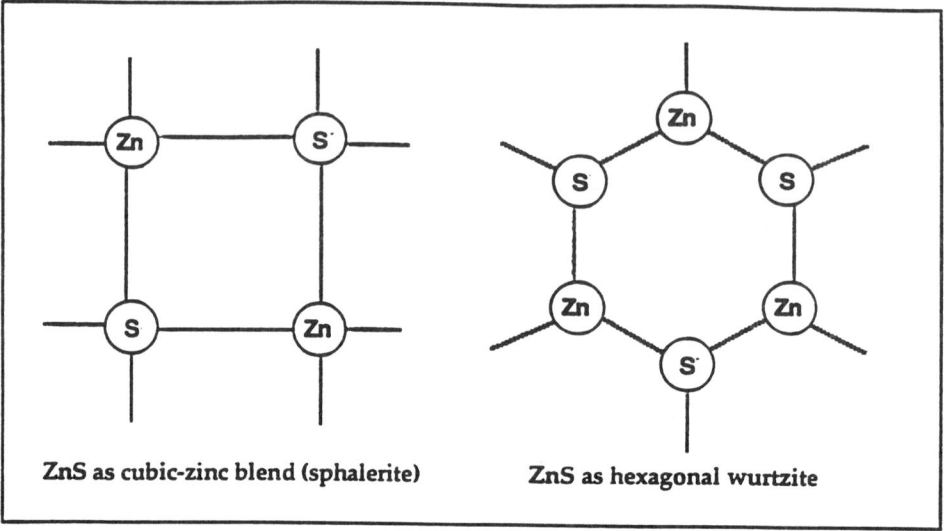
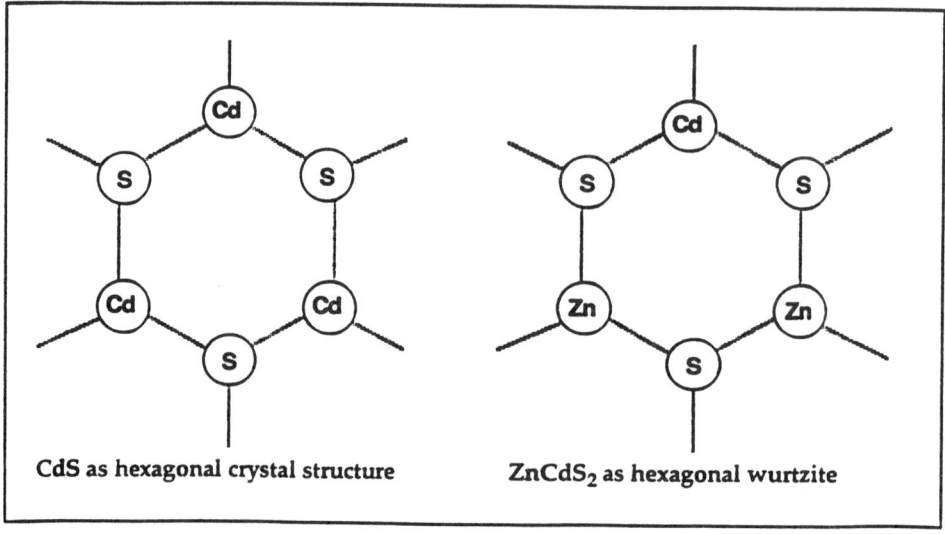

FIGURE 3-1. Structures of ZnS, CdS, and ZnCdS.

TOXICOKINETICS AND BIOAVAILABILITY OF ZNCDS: AVAILABILITY OF CADMIUM FROM ZNCDS

No studies on the toxicokinetics of ZnCdS were found. ZnCdS is insoluble in water and lipids and poorly soluble in strong acids. A small number of toxicity studies (which do not meet the current standards of toxicity testing) have suggested that it is not absorbed through the skin or gastrointestinal tract (Lawson and Alt 1965; Leighton and others 1965). The subcommittee believes that the lack of solubility of ZnCdS particles together with the limited toxicity studies implies that it will not be absorbed through the skin or gastrointestinal tract and that inhaled particles are not likely to be absorbed from the lung into blood for systemic distribution. Its lack of solubility also suggests that it is highly unlikely that free cadmium ions would become bioavailable to target organs as a result of inhalation of ZnCdS, However, information is not available on whether ZnCdS might break down in the respiratory tract into more-soluble components, which could be easily absorbed into the body.

TOXICITY

The subcommittee reviewed all available toxicity data on ZnCdS from the Army's files and from the open literature. The toxicity database on ZnCdS is limited and consists of eye irritation and dermal toxicity studies, single-dose oral toxicity studies, and observations reported in a few exposed to high concentrations of the dust for 1-2 years. These data are summarized below.

Eye irritation from exposure to a phosphor mixture that consisted of 65.4% liquid cosmetic base (composition not specified) and 34.6% ($\approx$35%) ZnCdS (Lawson and Alt 1965; Leighton and others 1965) was examined by instilling 0.1 mL of the test mixture in the eyes of adult rabbits and then observing them at 24, 48, and 72 h. The results of the experiment indicated that ZnCdS has negligible eye-irritation properties (Lawson and Alt 1965).

Dermal toxicity resulting from exposure to ZnCdS was examined by applying the test mixture at 9.4 g/kg of body weight to 4 rabbits for 24 h.

The mixture was injected under a rubber sleeve fitted around the clipped trunks of the test animals. No toxic effects were noted, and there was no evidence of dermal irritation during the 3 wk of observation after the treatment (Leighton and others 1965).

Lawson (1966) and Lawson and Alt (1965) reported on the medical use of ZnCdS for skin painting as a diagnostic tool for cancer. The compound is a phosphor whose fluorescence increases with increases in temperature. The authors used material that was composed of 59% CdS and 41% ZnS with less than 0.05% silver, less than 0.0005% nickel, and traces of halide to detect the higher temperature of blood in veins that leave a cancerous area. The phosphor was mixed into the same water-soluble cosmetic base used in the Army's toxicity studies and painted on the skin over the area of concern. A warm subcutaneous vein leaving a cancer could be clearly displayed by exposing the painted skin to UV light and observing the fluorescence of the phosphor. The authors stated that the ZnCdS phosphor was "sufficiently insoluble to be physiologically inert."

The only oral studies conducted by the Army were single-dose toxicity experiments in which rats and dogs were fed the mixture. None of the animals died at the highest doses tested, which were 10 g/kg of body weight and 20 g/kg of body weight for dogs and rats, respectively. These tests therefore indicated that the LD_{50} of the mixture—the dose that is lethal to 50% of the exposed animals—for dogs and rats was greater than 10 g/kg of body weight and 20 g/kg of body weight, respectively. Because the phosphor mixture used in the LD_{50} study contained 65.4% liquid cosmetic base and about 35% ZnCdS, the highest dose of the mixture tested in dogs and rats contained ZnCdS at 3.5 and 7.0 g/kg of body weight, respectively. To avoid physical injury from the administration of massive doses, higher doses were not administered to the animals. Thus it appears from these data that ZnCdS is not acutely toxic when given orally; that finding is consistent with the insolubility of the compound and its suspected lack of bioavailability.

No toxicity experiments of inhaled ZnCdS are available in the literature.

Because the ZnCdS particles used in the Army's dispersion studies were so small, the particles could probably be inhaled and deposited in the deep lung. The lack of solubility of the particles suggests that they are not likely to be absorbed from the lung into the blood for systemic distribution.

No information is available on the potential toxicity of the particles in the lung. It is also not known whether ZnCdS can be broken down by pulmonary macrophages into more-soluble forms of cadmium.

The Arkansas Department of Health evaluated the possible adverse effects of ZnCdS aerosol exposure from the Army's tests in White County, Arkansas (White 1977). The tests consisted of 131 releases of ZnCdS in 1967-1968. The evaluation was based on examining 4 workers who disseminated the chemical and were exposed to high concentrations of the dust for 1-2 yr, and examining data from the Arkansas Cancer Registry for any increases in incidence of lung, renal, or prostatic cancer from 1970 to 1975. There was no indication of cadmium-induced illness either in the 4 workers or in the general population. However, it should be noted that the followup period was not long enough to detect chronic effects, such as cancer. No studies examining the chronic toxicity, including cancer, of ZnCdS were found.

IMPLICATIONS OF VARIABLE COMPOSITION

The Army asked that the National Research Council assess the toxicologic implications of the variable composition of ZnCdS used in the Army's tests. The subcommittee has reviewed the available information on the composition of ZnCdS used in the Army's tests and found no evidence of substantial variation in the amounts of zinc and cadmium in the sintered compound. Any slight differences in composition should be reflected by equally small changes in potential toxicity.

CONCLUSIONS AND RECOMMENDATIONS

Data on the toxicity of ZnCdS are sparse. Results of feeding studies conducted in dogs and rats suggest that ZnCdS is practically nontoxic in acute (single, high-dose) exposures. ZnCdS was not found to be a skin or eye irritant in rabbits.

The physical and chemical properties of ZnCdS are fairly well understood. It is soluble only in strong acids and probably not absorbed through the skin or gastrointestinal tract. Its lack of solubility suggests that it is highly unlikely that free cadmium ions would become bioavailable as a result of inhalation of ZnCdS; however, information is not available on whether ZnCdS might break down in the respiratory tract into more-soluble components, which could easily be absorbed into the blood. This information and the limited toxicity data available suggests that ZnCdS is not likely to show systemic toxicity.

The subcommittee in its interim report (NRC 1995) recommended that the Army conduct studies to determine the bioavailability and inhalation toxicity of ZnCdS in experimental animals. This research will strengthen the database needed for assessing the risks of ZnCdS and lessen the need to rely on the use of a surrogate for toxicity information. The Army has begun the research recommended by the subcommittee, and the results are expected shortly. The subcommittee recommends that when the results of the research become available, they be reviewed by experts outside the Army to determine whether the subcommittee's conclusions are still valid or should be modified.

4

TOXICITY AND RELATED DATA ON SELECTED CADMIUM COMPOUNDS

FACED WITH THE TASK of evaluating the potential toxicity of zinc cadmium sulfide (ZnCdS), a compound with largely unknown toxic potential but reasonably well-known physical and chemical properties, the subcommittee considered it to be prudent to examine toxicity and related data on the most-toxic element in ZnCdS, cadmium. The toxicity of zinc, its interaction with cadmium, and the toxicity of copper and silver are discussed in Appendix D. This chapter reviews the physical and chemical properties, toxicokinetics, and toxicity of cadmium and cadmium compounds.

It should be noted that a substance can be insoluble in water or acids in vitro but soluble in vivo; cadmium oxide (CdO) is an example. (In vivo solubility is the ability of a material to leave the site of administration and be distributed systemically to other parts of the body.) The toxic potency of cadmium compounds depends on their in vivo solubility and bioavailability. The greater the solubility of the cadmium compound, the greater its systemic toxic potency. However, it is less clear whether water solubility is a major factor in determining the carcinogenicity of cadmium compounds. The bioavailability of a compound relies heavily, if not exclusively, on its physical and chemical properties, particularly its solubility. The bioavailability of a metal ion of a metal compound is a key determinant of the toxicity of the compound. This must be taken into account whenever an attempt is made to extrapolate the quantitative and qualitative effects of a metal compound with, say, high solubility and known toxicity to a compound with unknown toxicity, such as ZnCdS that contains the same metal but is poorly or not at all soluble in water or acid. Because ZnCdS is neither water-soluble nor apparently bioavailable, the subcommittee believes that the use of toxicity data on cadmium compounds to estimate the toxicity of ZnCdS constitutes a worst-case-scenario; this approach will lead to an overestimate of the risk associated with ZnCdS. The Army, in its risk-assessment reports on ZnCdS exposures, considered cadmium as a surrogate. Appendix G contains, in more detail, the subcommittee's evaluation of this approach. The Environmental Protection Agency, Agency for Toxic Substances and Disease Registry (ATSDR), and the Center for Disease Control reviewed the Army's risk-assessment reports; and Appendix H contains the subcommittee's review of those evaluations.

PHYSICAL AND CHEMICAL PROPERTIES OF CADMIUM COMPOUNDS

The metal cadmium is insoluble in water (Weast 1985). Cadmium sulfide, CdS, has low solubility; its reported solubility limit is 1.3 mg/L (9.0 mol/m^3) at 20°C (Weast 1985). The sulfate and chloride salts of cadmium have very high solubilities in water (Weast 1985): cadmium chloride, $CdCl_2$, has a reported solubility limit of 1,680,000 mg/L at 20°C and cadmium sulfate, $CdSO_4$, a reported solubility limit of 608,000 mg/L at 20°C. Table 4-1 shows the water-solubility ranges of selected cadmium compounds.

It is generally assumed that the toxicity of metal compounds usually involves an interaction between the free metal ion and a target tissue (Goyer 1995). Inorganic metal compounds that are soluble in water or can be made soluble in biologic fluids or in cells (such as alveolar macrophages) are typically more toxic than are metal compounds with poor solubility. Cadmium compounds are no exception. They have a wide range of water solubility, from 1.3 mg/L for CdS to 1,680,000 mg/L for $CdCl_2$ (see Table 4-1). The ATSDR toxicological profile on cadmium shows that cadmium compounds highly soluble in vivo (such as $CdCl_2$, CdSO, and cadmium oxide, CdO) are considerably more toxic and can cause death at substantially lower concentrations than cadmium compounds poorly soluble in vivo (such as CdS and cadmium carbonate, $CdCO_3$).

The only oxidation state of importance for cadmium under environmental conditions is +2. In the presence of sulfide ions and under reducing conditions, CdS is formed over a wide pH range. In aqueous systems, water hardness and pH determine the speciation of cadmium. In basic systems, cadmium hydroxide, $Cd(OH)_2$, can precipitate. In fresh water at typical environmental pH of 6-8, Cd^{2+} is the predominant species (Bodek and others 1988). The resulting precipitation of CdS in reducing environments can control the effective solubility of cadmium in natural waters.

TABLE 4-1 Water Solubilities of Selected Cd Compounds

Name	Formula	Water Solubility at 20°C
Cadmium metal	Cd	Insoluble
Cadmium chloride	$CdCl_2$	1,680,000 mg/L
Cadmium oxide	CdO	5 mg/L, but soluble in lung
Cadmium sulfide	CdS	1.3 mg/L
Cadmium carbonate	$CdCO_3$	Insoluble
Cadmium red	70% Cd, 16% Se, 13% S	Insoluble
Cadmium yellow	77% Cd, 22% S, 0.3 % Zn	Insoluble
Cadmium selenide	CdSe	Insoluble
Cadmium sulfate	$CdSO_4$	608,000 mg/L
Zinc Cadmium Sulfide	ZnCdS	Insoluble

Source: Adapted from Weast 1985.

TOXICOKINETICS OF CADMIUM COMPOUNDS

INSOLUBLE CADMIUM COMPOUNDS

Klimisch (1993) studied the lung deposition, lung clearance, and renal accumulation of inhaled CdS in rats. Rats were exposed to CdS at 0.2, 1.0, and 8.0 mg/m^3 6 h/day for 10 d. About 40% of the lung burden was cleared rapidly ($t_{1/2}$ = 1.4 d) and 40% slowly ($t_{1/2}$ = 42 d), leaving a residue of 20% of the initial lung burden 90 d after the end of the exposure. Only 1% of the CdS cleared from the lungs accumulated in the kidneys. Klimisch then used the data of Loeser (1974) on the amount of cadmium accumulated in the kidneys after oral exposures to CdS to calculate an intestinal-absorption factor of 0.02% for CdS. The absorption factor for inhaled CdS based on the data of Rusch and others (1986) was 0.07–0.1%. That contrasts with results of similar studies with $CdCl_2$ (which is water-soluble) in which 35% of the material cleared the lungs and accumulated in the kidneys. The author concluded that the bioavailability of cadmium from CdS is much lower than that from the more-soluble forms of cadmium, such as $CdCl_2$.

A study on the toxicokinetics of inhaled CdS (Oberdörster 1990; Oberdörster and Cox 1990) provides evidence that inhaled CdS is not absorbed from the lungs into systemic circulation. Rats were given CdS by inhalation. Failure to detect cadmium in their kidneys, the primary target organs for cadmium distribution and toxicity, suggested that the cadmium in the inhaled particles did not leave the lungs, enter the bloodstream, and affect the kidneys. No inflammatory response to CdS was detected in the lung after exposure to the same amount of substance as had caused an inflammatory response to $CdCl_2$ and CdO. Of the inhaled CdS, 75% was exhaled immediately. Oberdörster (1990) estimated that 90% of the remaining inhaled CdS was removed slowly from the lung into the gastrointestinal tract by normal mechanical clearance processes. Rusch and others (1986) exposed rats for 2 h to various cadmium compounds at 100 mg/m^3 (based on cadmium content) and found that insoluble forms of cadmium, such as cadmium red and cadmium yellow (furnace-treated dyes that contain cadmium in the form of CdS and cadmium selenide, CdSe), were not transported from lung to kidney. For example, cadmium yellow consists of crystals containing 77% cadmium, 22% sulfur, and less than 1% zinc and selenium; after inhalation exposure at high concentrations, kidney cadmium content was indistinguishable from that in controls.

SOLUBLE CADMIUM COMPOUNDS

Soluble cadmium compounds can be absorbed from the skin, intestinal tract, or respiratory tract into the bloodstream and transported throughout the body with the potential for causing systemic injury. When such compounds are absorbed, the most-sensitive sites for injury are the lungs, kidneys, and bones; studies have also been conducted on the effects of soluble cadmium compounds on the immune system and the reproductive and developmental system.

TOXICITY OF CADMIUM COMPOUNDS

NONCANCER EFFECTS

The subcommittee reviewed the toxicity of in vivo soluble and insoluble cadmium compounds. Several comprehensive reviews on the toxicity of cadmium compounds are available (ATSDR 1993; IARC 1993; OSHA 1992; IPCS 1992). The main focus of the subcommittee's review was to evaluate the toxicity associated with acute inhalation exposures to cadmium, because the exposures from the use of ZnCdS as a tracer particle were brief and sporadic. These data are summarized below.

IN VIVO INSOLUBLE CADMIUM COMPOUNDS

Many animal studies have identified toxic responses to the inhalation or ingestion of soluble cadmium compounds, but relatively few studies have examined the health effects of cadmium compounds that are poorly soluble in vivo, such as CdS, CdSe, $CdCO_3$, or cadmium dyes. Nevertheless, enough experimental information is available to show that the toxicity of

CdS is quite different from the toxicity of soluble cadmium compounds. The main differences are that it takes considerably higher CdS concentrations to produce lung lesions and that substantially less cadmium becomes bioavailable from inhaled or ingested CdS than from the more-soluble cadmium compounds.

Studies that compared the potential inflammatory response to intratracheal instillation of 30 µg of different cadmium compounds in rats ($CdCl_2$, CdO, and CdS) reported that CdS failed to elicit any substantial influx of neutrophils, lavagable protein and enzyme activities in bronchoalveolar lavage fluid 24 h after exposure, whereas the other cadmium compounds tested resulted in substantial changes in these characteristics (Oberdörster and others 1985).

When rats inhaled CdS at 1,000 µg/m³ 22 h/d, 7 d/wk for 30 d, a mild inflammatory response was observed in lungs, but recovery had occurred 2 mo after exposure (Glaser and others 1986).

In Vivo Soluble Cadmium Compounds

Lung—In humans, high concentrations of CdO fumes or dusts (milligrams per cubic meter) are irritating to the respiratory tract; irritation does not occur following lower inhalation exposures. (CdO is insoluble in water; however, it is soluble in vivo.) Long-term chronic occupational exposure can cause emphysema in humans. In experimental animals, inhalation of soluble cadmium compounds has been shown to produce acute lung injury (chemical pneumonitis) and chronic lung injury (fibrosis and emphysema).

In an inhalation study by Buckley and Bassett (1987), the lowest-observed-adverse-effect level (LOAEL) resulting in mild, reversible pulmonary inflammation in rats exposed to CdO was 500 µg/m³ (the equivalent of cadmium at about 440 µg/m³) following a 3-h exposure. In a similar study by Grose and others (1987), the highest air concentration of $CdCl_2$ or CdO that did not result in detectable lung lesions in rats and rabbits exposed for 2 h was the equivalent of cadmium at 450 µg/m³; this concentration was considered a no-observed-adverse-effect level (NOAEL) (ATSDR 1993). Thus, for 2 h, an exposure at 450 µg/m³ (900 µg-hr/m^3) was a NOAEL in the studies by Grose and others (1987), and exposure for 3 h at 440 µg/m^3 (1,320 µg-hr/m^3) was a LOAEL in the Buckley and Bassett study (1987). The two studies clearly define that for inhalation exposure to soluble cadmium compounds, the NOAEL for the respiratory tract in the rat is cadmium at 900-1,300 µg-hr/m^3.

KIDNEY—The kidney is a major target organ for cadmium toxicity after chronic, low-level exposure to forms of cadmium that can be absorbed by the body and transported to the kidney. Workers occupationally exposed to cadmium and cadmium compounds (for example, CdO dust and cadmium fumes in factories producing nickel-cadmium batteries) suffer from a high incidence of abnormal renal function, indicated by proteinuria and a decrease in glomerular filtration rate (Falck and others 1983; Friberg 1950; Thun and others 1989). (For an extensive review of the literature on effects of cadmium on the kidney, see ATSDR 1993.)

An exposure to cadmium that produces a kidney concentration of 200 µg/g wet weight is considered a threshold for renal dysfunction in an adult population chronically exposed to cadmium (Friberg and others 1974; Kjellstrom and others 1977, 1984; Roels and others 1983). Cadmium concentrations in the kidneys of North American adults are about 20 µg/g wet weight in nonsmokers and 40 µg/g in smokers (Chung and others 1986).

ATSDR (1993) has compiled a list of animal studies whose results yield NOAELs for renal effects of cadmium exposure. The NOAELs for acute exposure, which is the type of exposure of concern to the subcommittee, were considered for inhalation and oral routes. No acute-inhalation studies of cadmium toxicity to the kidney were listed; acute oral exposures to CdCl$_2$ as high as 150 mg/kg for 1 d in rats were reported to have no renal effect (Kotsonis and Klaassen 1977). Chronic exposure of humans at 0.0021 mg/kg per day for a lifetime has been considered to have no renal effect (Nogawa and others 1989). ATSDR has set a chronic inhalation minimal-risk level (defined as an estimate of the highest daily human exposure to a chemical at a dose that is likely to be without appreciable risk of adverse noncancer effects over a specified duration of exposure) for cadmium-induced renal toxicity of 0.2 µg/m^3 (ATSDR 1993). OSHA's estimates of risk posed by exposure to a time weighted average (TWA) of 5 µg/m^3 range between 14 and 23 excess cases of kidney dysfunction per 1,000 workers (OSHA 1992).

BONE—Cadmium affects calcium metabolism by decreasing the rate of bone formation and increasing calcium loss from bone. The relevant studies have been reviewed by Bhattacharyya and others (1995). The effects of low-dose exposures on bone metabolism are subtle, but chronic exposures, particularly in combination with other factors, lead to calcium loss from bone. Factors contributing to the severity of cadmium toxicity to bone include renal disease, dietary deficiencies (particularly low dietary calcium), vitamin D deficiency, and low estrogen concentrations. As in renal toxicity, cadmium toxicity to bone results from exposure to forms of cadmium that are soluble enough to allow it to reach the bone. No published studies have shown bone loss from short-term high-level exposures to cadmium.

IMMUNE SYSTEM—The effects of cadmium on the immune system of animals or humans are inconsistent (Cifone and others 1990; Exon and Koller 1986; Funkhouser and others 1994; Horiguchi and others 1993; Kastelan and others 1981; Koller 1973), and the in vitro effects do not correlate well with the in vivo effects. Cadmium stimulated the immune response in some studies and suppressed it in others. A NOAEL cannot be determined until a definite immune pattern is characterized for this chemical.

DEVELOPMENTAL AND REPRODUCTIVE SYSTEM—ATSDR has published a toxicological profile on cadmium (ATSDR 1993) that includes an excellent review of the reproductive and developmental toxicity of cadmium. We will not reiterate all of the reported studies but highlight here the ones most relevant for this report. It should be noted that most studies have focused on occupational exposure of male workers to cadmium. Noticeably absent are environmental studies of cadmium exposure that include potentially sensitive populations, such as pregnant or nursing women, children, and the elderly. Of most interest to the subcommittee were reports of the effect of acute exposures to cadmium on the developmental and reproductive systems, but such studies were not available.

DEVELOPMENTAL TOXICITY—Two epidemiologic studies looked at pregnancy outcomes in occupationally exposed populations (ATSDR 1993); the primary route of exposure was probably inhalation. A Russian study observed lower birthweights of children in cadmium-exposed women (exposures up to 35 mg/m^3) than in unexposed controls; however, the decrease was not correlated with duration of maternal exposure to cadmium (Tsvetkova 1970). Dose-response information was not available for this study. A study of French women occupationally exposed to heavy metals showed doubled concentrations of cadmium in the hair of mothers and newborns relative to matched controls (1.45 versus 0.59 ppm in exposed mother's hair and 1.27 versus 0.53 ppm in transplacentally exposed newborn hair) but did not show any adverse effects in the newborns (Huel and others 1984).

Loiacono and others (1992) conducted a prospective study of women living around a lead smelter plant in Yugoslavia. Women were believed to have been exposed to lead and cadmium dust from the smelter. The study's hypothesis was that placental cadmium accumulation was associated with decreased birthweight. Multiple regression was used to control for potential confounders of birthweight and gestation, such as smoking; no association was found between placental cadmium and birthweight or gestational age.

Cadmium has been shown to cause developmental toxicity in rodents after inhalation. Such effects as delay in ossification, decrease in locomotor activity, and impairment of reflexes were noted after exposures to CdO as low as 0.02 mg/m^3, 5 h/d 5 d/wk for 5 mo before mating and from the 1st to the 20th day of pregnancy (Baranski 1985). No adverse effects on maternal homeostasis were reported for this exposure, but maternal toxicity was significant at the two higher exposures tested (CdO at 0.16 and 1.0 mg/m^3). Therefore, one can infer from this study that the ambient concentrations of CdO might cause developmental toxicity in humans. However, CdO is soluble in vivo whereas ZnCdS is insoluble in vivo; therefore, the results of Baranski (1985) cannot be extrapolated to inhalation exposures involving ZnCdS. After exposures of female pregnant rats to CdO at 0.16 mg/m^3, decreases in weight gain, osteogenesis, and viability were also observed in offspring. Both maternal and fetal weight changes were noted after exposure to CdCl$_2$ for 21 d at 0.6 mg/m^3 (Prigge 1978). Numerous rodent studies have shown that oral exposure to soluble cadmium can reduce fetal or pup weights and cause skeletal malformations (ATSDR 1993). Oral exposure to cadmium compounds soluble in water has been shown to cause reduced fetal or pup weights (ATSDR 1993; Ali and others 1986; Baranski 1987; Kelman and others 1978; Petering and others 1979; Pond and Walker 1975; Sorell and Graziano 1990; Sutou and others 1980; Webster 1978; Whelton and others 1988). A few of those studies have also shown malformations, especially of the skeleton (Baranski 1985; Machemer and Lorke 1981; Schroeder and Mitchener 1971). Neurodevelopmental effects in rats have been identified as the most sensitive end point in these oral-exposure studies with cadmium compounds soluble in water. Baranski and others (1983) reported reduced locomotor activity and impaired balance in pups from dams exposed to 0.04 mg/kg per day before and throughout gestation, and Ali and others (1986) reported hyperactivity in offspring from dams exposed to cadmium at 0.7 mg/kg per day during gestation. Chisolm and Handorf (1985, 1987) have also postulated that cadmium exposure can play a role in pregnancy-induced hypertension and suggested that animals late in gestation might be uniquely sensitive to this effect. Those studies suggest that the ambient concentrations of CdO, a compound soluble in vivo, might cause developmental toxicity in humans. However, ZnCdS is an insoluble cadmium compound, and that suggests that it is unlikely to produce developmental toxicity at the low ambient concentrations. (See discussion earlier in this chapter about the solubility of cadmium compounds and implications for noncancer end points.)

REPRODUCTIVE TOXICITY—Early studies of the effect of cadmium exposure on human male fertility did not find adverse effects on fertility (Kazantzis and others 1963) or on urinary androgen excretion (Favino and others 1968). However, those early studies included only 12 and 10 occupationally exposed men, respectively, so the findings need to be viewed as preliminary.

Limited negative evidence from two occupational epidemiologic studies suggests that inhalation exposure to low doses of cadmium does not cause serious reproductive effects in men or women (Mason 1990; Tsvetkova 1970). Only a few end points were evaluated in those studies including menstrual cycling and serum hormone concentrations. Such important end points as fertility and reproductive function were not studied. Cadmium has been shown to concentrate in testicular tissue after occupational exposure (Smith and others 1960).

Saaranen and others (1989) compared 24 men attending fertility clinics with 38 fertile men whose wives conceived within 6 mo to determine whether cadmium concentration in seminal fluid and sperm affected fertility. Concentrations of lead, cadmium, and zinc in the testes of 41 autopsied men were evaluated (Oldeneid and others 1993); no associations between seminal-plasma cadmium content and semen quality or fertility changes were noted, despite the ability of cadmium to reach human semen. Favino and others (1968) did not detect changes in androgen function in 10 workers in a cadmium-alkaline battery plant even though measurable amounts of urinary cadmium were detected and workers had experienced ulceration of the nasal mucosa and anosmia.

DERMAL TOXICITY—A search of the available reports on the toxicity of cadmium and its compounds failed to identify any studies in which skin exposures have produced serious adverse health effects in humans (ATSDR 1993). In a few studies, patients were exposed to dermally applied solutions of 2% $CdCl_2$ in distilled water. Of 1,502 patients tested, only 25 showed a positive reaction that was interpreted as reflecting acute skin irritation rather than contact allergy to cadmium (Wahlberg 1977). One patient showed a reaction also when tested with a 1% $CdCl_2$ solution. However, in another study involving close to 1,500 patients, a 1% $CdCl_2$ solution caused no skin toxicity (Rudzki and others 1988).

OTHER NONCANCER TOXICITY— Cardiovascular, gastrointestinal, hematologic, musculoskeletal, or hepatic effects in humans or experimental animals following exposure to cadmium have not been reported (ATSDR 1993).

GENOTOXICITY OF CADMIUM COMPOUNDS

This section reviews the genotoxicity of soluble and insoluble cadmium compounds.

A wide range of tests have been conducted to determine the genotoxicity of cadmium. It has been tested in bacteria, plants, insects, and mammalian cells, including human cells, in vitro and in vivo. Comprehensive reviews of the various investigations have been provided by Occupational Safety and Health Administration (OSHA 1992) and the Agency of Toxic Substances and Disease Registry (ATSDR). Both positive and negative results were reported. Positive mutagenicity results were generally observed with soluble cadmium compounds (such as $CdCl_2$), whereas insoluble compounds generally yielded negative results.

Positive mutagenicity were observed in some studies that used bacterial cells (Bruce and Heddle 1979; Kanematsu and others 1980; Mandel and Ryser 1984; Wong 1988) and in most studies that used yeast or mammalian cell cultures (Denizeau and Marion 1989; Oberly and others 1982). Chromosomal aberrations have been found in most studies that involved cadmium treatment of mammalian cells (Deaven and Campbell 1980; Rohr and Bauchinger 1976) and in some studies that used human lymphocytes in culture (Gasiorek and Bauchinger 1981; Shiraishi and others 1972) and bone marrow cells after intraperitoneal (Mukherjee and others 1988a) and oral (Mukherjee and others 1988b) exposure of mice.

Evidence of chromosomal aberrations in humans after inhalation (Deknudt and Leonard 1975; O'Riordan and others 1978) or oral (Bui and others 1975; Tang and others 1990) exposure to cadmium compounds is conflicting. Cadmium does not appear to cause germ-cell mutations or chromosomal damage after oral (Sutou and others 1980; Zenic and others 1982) or intraperitoneal (Epstein and others 1972; Mailhes and others 1988) exposure in animals, but does so after subcutaneous exposure (Watanabe and Endo 1982; Watanabe and others 1979).

Overall, cadmium appears to have the capability of altering genetic material, particularly chromosomes in mammalian cells, but germ cells appear to be protected except at high acute parenteral doses (ATSDR 1993). The subcommittee concludes that until more conclusive mutagenicity studies are conducted and reported, cadmium may be considered to be a potential mutagenic agent.

CARCINOGENICITY OF CADMIUM COMPOUNDS

Studies on the carcinogenicity of cadmium and cadmium compounds were reviewed most recently by the International Agency for Research on Cancer (IARC) in 1993. IARC concluded that cadmium and cadmium compounds are carcinogenic in humans on the basis of epidemiologic studies of occupationally exposed workers, experimental studies in animals, and genotoxic effects in a variety of types of eukaryotic cells, including human cells (IARC 1993). A less-complete review by ATSDR in the same year characterized the evidence of carcinogenicity in animals as "strong" but of that in humans as "limited" (ATSDR 1993). Most of the evidence of carcinogenicity involves lung cancer after long-term inhalation.

LUNG CANCER

Epidemiologic studies show some increase in lung cancer among workers in 5 of 6 occupationally exposed populations in Europe and the United States (IARC 1993). Workers were exposed predominantly to cadmium as CdO (fume or dust) or CdS. The human evidence linking cadmium to lung cancer is strongest in the U.S. plant where exposures were well characterized, and the persistent dose-response relationship with cadmium cannot be readily explained by cigarette-smoking or other known occupational lung carcinogens (IARC 1993).

Epidemiologic studies of cancer in workers exposed to cadmium have been conducted in five plants in the United States, England, and Sweden (Tables 4-2 and 4-3). Two of the plants (groups 1 and 2) are metallurgic facilities, two (groups 4 and 5) are nickel-cadmium battery manufacturers, and one (group 3) is a compendium of 17 plants in the U.K., including a large zinc smelter. The mortality experience of these populations has been studied repeatedly (Elinder and others, 1985); Tables 4-2 and 4-3 present the most recent results. A statistically significant increase in mortality from lung cancer has been reported among U.S. cadmium-recovery workers with 2 or more years of employment (Thun and others 1985, Thun 1990), in nickel-cadmium battery workers in the U.K. (Sorahan 1987) and Sweden (Elinder and others 1985; Jarup and others 1990), and in the British 17-plant study (Kazantzis and Armstrong 1982; Kazantzis and others 1988).

TABLE 4-2 Lung-Cancer Mortality Among Cadmium Workers

Group	Type of Industry	Deaths Observed/ Expected	Standardized Mortality Ratio	95% Confidence Interval
1	Cadmium recovery plant, U.S. (Thun and others 1990)			
	—Overall cohort	24/17.50	137	(88-204)
	—2+ yr	24/10.76	223	(143-332)
2	Cadmium-copper alloy plant, U.K. (Kazantzis and Armstrong 1982)	47/46.4	101	(72-130)
3	17 plants, U.K. (Kazantzis and others 1988)	277/240.9	115	(101-129)
4	Nickel-cadmium battery, U.K. (Sorahan 1987)	110/84.5	130	(107-157)
5	Nickel-cadmium battery, Sweden (Jarup and others 1990)	14/5.8	241	(132-405)[a]

[a]20 years of employment, compared with regional rates.

Table 4-3 indicates those cohorts for which a dose-response relationship is present between cadmium exposure and mortality from lung cancer. In three cohorts (Thun and others 1990; Sorahan 1987; Kazantzis and others 1988), the standardized mortality ratio (SMR) for lung cancer increases with either length of employment or cumulative exposure to cadmium.

TABLE 4-3 Dose-Response Relationships for Lung Cancer in Studies of Cadmium-Exposed Workers

Group	Type of Industry	Dose-Response Relationship Evident	Range in Standardized Mortality Ratio
1	Cadmium recovery plant, U.S. (Thun and others 1990)	Yes	46, 263, 373[a]
2	Cadmium-copper alloy plant, U.K. (Kazantzis and Armstrong 1982)	No	87, 114, 72
3	17 plants, U.K. (Kazantzis and others 1988)	Yes	112, 121, 194[b]
4	Nickel-cadmium battery, U.K. (Sorahan 1987)	Possibly	(Marginal trend before 1947)[c]
5	Nickel-cadmium battery, Sweden (Jarup and others 1990)	No	234, 232[d]

[a]Death rates in cadmium workers compared with men in Colorado, followup to 1984.
[b]Author does not attribute to cadmium.
[c]Latency not considered.
[d]Regional rates, 20-yr latency.

Despite the value of the occupational studies in providing information about the potential carcinogenicity of cadmium, there are major differences in the intensity and duration of exposure between the communities exposed to ZnCdS by the Army and the occupational conditions. For example, at the U.S. cadmium-recovery plant, air concentrations of cadmium averaged 1,000 $\mu g/m^3$ before 1950 and 500 $\mu g/m^3$ from 1950 to 1960. Those measures are average concentrations breathed over an 8-h workday, often for decades. Because of the high cumulative exposures to cadmium, workers at the U.S. plant often developed cadmium-induced kidney toxicity (low-molecular-weight proteinuria) and obstructive lung disease even without smoking (Thun and others 1985).

CdS can produce lung tumors in rats. Rats exposed to 10 weekly intratracheal instillations of 250 μg of CdS had a statistically significant increase in the incidence of lung tumors (OSHA 1992; Pott and others 1987). In inhalation studies, a statistically significant increase in number of lung tumors (bronchioalveolar, adenocarcinoma, and squamous cell tumors) was seen in rats exposed 22 h/day, 7 d/wk for 18 mo at concentrations of 90 $\mu g/m^3$ or higher (Oldiges and others 1989). Later studies by König and others (1992) indicated that 50-63% of CdS used in the inhalation studies of Oldiges and others (1989) might have been converted to the more-soluble $CdSO_4$, which is known to be a pulmonary carcinogen in rats. Heinrich and others (1989), who exposed both mice and hamsters by inhalation to CdS at concentrations of 90-1,000 $\mu g/m^3$ for up to 64 wk, failed to identify any statistically significant increase in incidence of lung tumors but did report increases in lipoproteinosis, fibrosis, and hyperplasia. A better understanding of the molecular and cellular mechanisms of cadmium carcinogenicity is required to determine how different the carcinogenic potential of CdS is from that of more-soluble cadmium compounds.

Human data from studies of occupationally exposed workers have been used to estimate cancer risk associated with exposure to cadmium (OSHA 1992). On the basis of simple linear regression, lung-cancer risk is estimated to increase with cumulative cadmium exposure by 14.6×10^{-6} per $\mu g/m^3$. That estimate is used in risk calculation in Chapter 6.

PROSTATIC CANCER

Epidemiologic studies of cadmium and prostatic cancer are inconclusive. Early studies of occupational exposure suggested increased deaths from prostatic cancer, but the association weakened with continued followup and has not shown a clear dose-response trend (IARC 1993). Japanese men who ingested cadmium-contaminated rice during World War II experienced higher mortality from prostatic cancer (standardized mortality ratio, 1.66) and a higher incidence of prostatic hyperplasia in one polluted area but not in another (Shigematsu and others 1982). The high prevalence of cadmium-induced kidney toxicity (30-80%) in both areas indicated systemic circulation of cadmium (IPCS 1992). Animal experiments do not show that inhaled or ingested cadmium induced prostatic cancer.

OTHER CANCERS

Experimental studies of animals have not consistently shown other types of cancer after cadmium inhalation or ingestion (IARC 1993). Leukemia can be produced by cadmium ingestion in zinc-deficient rats; it can be prevented by simultaneous zinc supplementation (IARC 1993). Epidemiologic studies in the Toyama prefecture of Japan, where residents ate highly contaminated rice, found no evidence of increased death from stomach or liver cancer or all cancers combined (Shigematsu and others 1982). No increase in other cancers about which concern was expressed in the public meetings on ZnCdS has been observed in the highly exposed communities in Japan. (IPCS 1992).

CONCLUSIONS

Faced with the task of evaluating the potential toxicity of ZnCdS, a substance with largely unknown toxic potential but reasonably well-known physical and chemical properties, the subcommittee considered it to be prudent to examine the toxicity and related data on the most-toxic element in ZnCdS, cadmium. On the basis of a review of the physical and chemical properties, toxicokinetics, and the toxicity of cadmium compounds, the subcommittee reached the following conclusions:

• Bioavailability of cadmium is a key determinant of the toxicity of a cadmium compound. Cadmium compounds differ in their toxicity: compounds that are insoluble in vivo are less toxic than soluble ones. Inhaled particles of ZnCdS would have direct contact with lung tissue but minimal systemic absorption due to poor solubility. Its lack of solubility also suggests that it is highly unlikely that free cadmium ions would become bioavailable to target organs as a result of inhalation of ZnCdS.
• Toxicokinetic studies with CdS showed that it is much less bioavailable than cadmium compounds that are soluble in vivo.
• Cadmium compounds that are soluble in vivo are toxic primarily to the lungs, kidneys, and bone. There are very few reported studies on the toxicity of cadmium compounds that are poorly soluble in vivo, such as

CdS. In several studies, CdS failed to elicit lung or other target-organ toxicity in experimental animals, presumably because it is not bioavailable.

- Because ZnCdS is neither water-soluble nor apparently bioavailable, the subcommittee believes that exposure to ZnCdS is not likely to produce adverse health effects and that the use of toxicity data on soluble cadmium compounds to estimate the potential for noncancer toxic effects of ZnCdS would constitute a worst-case scenario. Thus, this approach will overestimate the risk.
- Noncancer-toxicity data are available on soluble and insoluble cadmium compounds, but the only cancer-potency data available on cadmium are based on relatively high occupational exposures to cadmium compounds, mostly of CdO or mixtures of CdO and CdS.
- Inhaled cadmium has been shown in occupational studies and laboratory studies of animals to cause lung cancer but not cancer at other body sites.
- Cadmium exposures associated with increased lung-cancer risk in human and animal studies were to much higher concentrations for longer periods and involved more biologically soluble compounds than the exposures to cadmium from ZnCdS in the Army's testing program.
- A quantitative risk assessment for lung cancer based on occupational exposure of humans to cadmium compounds is likely to overestimate the lung-cancer risk for ZnCdS exposures from the Army's dispersion tests.

5

Exposure Assessment

THIS CHAPTER CONTAINS the subcommittee's exposure assessment for zinc cadmium sulfide (ZnCdS) and cadmium.

ZINC CADMIUM SULFIDE

The subcommittee reviewed the Army's sampling and analytic methods for assessing exposures to ZnCdS and the Army's documents on exposure to ZnCdS in various locations. The Army measured the concentration of ZnCdS particles in its dispersion studies with impingement and filtration methods. The methods are described in Appendix F. The subcommittee concludes that those are accurate and appropriate methods for measuring ZnCdS.

The subcommittee also reviewed the Army's documents and estimated the concentrations and potential exposures (time-integrated concentrations) of ZnCdS that were achieved during the use of the compound in the Army's air dispersion tests. Table 5-1 provides information on location, dates, numbers, and quantities of ZnCdS releases. It also provides the maximum concentrations and exposure doses (concentration × time). The

TABLE 5-1 Exposure Data on ZnCdS Dispersion Tests

Ref.	Place	Name	Start Date	End Date	No. of Releases	Total Quantity of ZnCdS, kg	Populated Areas Max. Exp. to ZnCdS, µg-min/m³	Approximate area affected, sq. miles	Max. Conc. of ZnCdS, µg/m³	Cadmium inhalation intake, µg
			Releases not at Dugway Proving Ground							
2	Camp Cooke, Calif.		1955		39	2.3	<173 <10		<7 <10	<0.4
8	N. Carolina, S. Carolina, Georgia,	DEW I	03/26/52	04/21/52	5	630	98 <100,000		0.34 <10,000	0.3
13	Corpus Christie, Tex.	WINDSOC	08/13/59	02/22/60	13	~1,600	NA <100,000		NA <10,000	NA
16	Oklahoma		06/04/62	06/16/62	9	204	39 <1,000		0.75 <1,000	0.1
16	Texas		06/24/62	06/29/69	9	204	36 <1,000		0.82 <1,000	<0.1
16	Washington		10/02/62	10/21/62	9	204	6.7 <1,000		0.3 <1,000	<0.1
16	Nevada		10/31/62	11/05/62	8	181	23 <1,000		1.6 <1,000	<0.1
17 & 43	St. Louis		05/27/63	03/17/65	42	984	7,400 <10		40 <10	19.2
19	Chippewa National Forest, Minn.		01/25/64	08/07/64	24	330	1,620 <1,000		3 <1,000	4.2
20	San Francisco		03/25/64	04/23/67	18	27.75	<1,900 <10		<170 <1	4.9
22	Fort Wayne, Ind.		02/02/64	02/04/66	75	~1,650	410 <1,000		<20 <1,000	1.1
24	Oceanside, Calif.	Onshore	06/23/67	07/17/67	45	237	694 <1		1,741 <1	1.8
		Offshore releases					149 <1,000		1.3 <1,000	0.4
27	Pack Forest, Wash.		10/15/68	09/05/69	33	1.7				?
28	Green Brier Swamp, Md.	MATE	08/01/69	10/29/69	111	2.7	<42 <1		~0.03 <1	<0.1

TABLE 5-1 (Continued)

							Populated Areas		
Ref.	Place	Name	Start Date	End Date	No. of Releases	Total Quantity of ZnCds, kg	Max. Exp. to ZnCdS, µg-min/m³ Approximate area affected, sq. miles	Max. Conc. of ZnCdS, µg/m³	Cadmium inhalation intake, µg
30	Camp Detrick, Md.	SELTZER	02/18/53	02/24/53	4	0.022	71 <10	9 <1	0.2
30	Biltmore Beach	WHITEHORSE	03/24/53	05/02/53	12	9.7	150,000 <0.1	4,800 <0.1	390
33	Dallas		04/01/61	08/31/61	37		NA <1,000	NA <1,000	NA
35	St. Louis		01/19/53	10/18/53	35		2,000 <10	340 <0.1	5.2
35	Minneapolis		05/20/53	06/23/53	102	7.9	2,600 <10	300 <0.1	6.8
35	Winnipeg		07/09/53	08/01/53	36	5.8	5,600 <10	1,000 <0.1	14.5
36	Stanford University, Calif.		10/15/47	10/15/47	1	0.00083	5	2	<0.1
37	Palo Alto, Calif.		03/10/50	03/14/50	2	0.976	2.4 <100	0.5 <100	<0.1
37	San Francisco		10/20/50	10/27/50	6	22.44	436 <100	15 <100	1.1
41	Palo Alto, Calif.		01/26/62	11/16/62	28	1.4	1,676 <0.1	? <0.1	4.4
			Releases at Dugway Proving Ground						
3	Dugway		05/04/53	06/03/53	2	5.534			
4	Dugway		01/21/54	03/14/54	4	0.0348			
6	Dugway		05/18/55	05/18/55	2	0.0424			
18	Dugway		05/17/63	08/15/63	9	29.6			
29	Dugway	GOOF	08/23/55	11/01/55	5	0.8	<409		<1.1
31	Dugway		04/03/58	04/22/58	4	0.0534			
32	Dugway		02/70 to 03/70		6	0.21			
37	Dugway		07/01/50	08/04/50	9	8.244	0.03		<0.1

four highest ZnCdS exposures were in Minneapolis (44 µg), Winnipeg (93 µg), St. Louis (156 µg), and Biltmore Beach, FL (2,500 µg). To obtain the exposure to cadmium, the maximal exposure of ZnCdS (expressed as µg-min/m^3 as shown in Table 5-1) is multiplied by 0.0166 m^3/min (the volume of air inhaled by an active person in 1 min). The product is then multiplied by 0.156 (the mass fraction of cadmium in ZnCdS). For example, the corresponding amounts of cadmium in the ZnCdS doses are 6.8 µg in Minnesota, 14.5 µg in Winnipeg, 24.4 µg in St. Louis, and 390 µg in Biltmore Beach.

CADMIUM

The purpose of this section is to provide estimates of the magnitude of potential cadmium doses from human contact with cadmium compounds as a result of the dispersion of ZnCdS by the Army and to compare them with the estimated doses from environmental and industrial sources. In the United States, mean concentrations of cadmium in ambient air range from less than 0.001 µg/m^3 in remote areas to 0.005-0.04 µg/m^3 in urban areas (Davidson and others 1985; Elinder 1985; EPA 1981; Saltzman and others 1985). Atmospheric concentrations of cadmium are generally highest in the vicinity of cadmium-emitting industries, such as smelters, municipal incinerators, and fossil-fuel combustion facilities. Measurements of atmospheric cadmium up to 7 µg/m^3 have been reported in these industrial types of areas in the United States (Schroeder and others 1987). Cadmium intake from air is estimated to be 0.1-0.8 µg per day in typical U.S. urban areas and less than 0.02 µg in rural areas.

Food is the largest potential source of cadmium exposure for the general population. There are several estimates of the daily adult intake of cadmium from food in the United States, but there is considerable variation among those estimates. Schroeder and Balassa (1961) reported a range of 4-60 µg and Nriagu (1981) a range of 38-92 µg, whereas estimated daily averages have been reported to be 30 µg (Gartrell and others 1986), 38 µg (Duggan and Corneliussen 1972), 50 µg (Duggan and Corneliussen 1972), 51 µg (Mahaffey and others 1975), and 92 µg (Murthy and others 1971). A more recent estimate based on a total diet study shows the daily dietary intake to be about 15 μg (Gunderson, 1995). Analysis of the earlier data shows that these discrepancies are probably due to different analytic methods. Cadmium contamination of food has been reduced over the years, presumably because of better technology. However, the cadmium contamination encountered in the 1950s and 1960s, when the Army's dispersion tests were conducted, are more relevant for risk assessment. On the basis of the U.S. data and data from other industrial nations in the Northern Hemisphere, the subcommittee believes that the daily cadmium intake from food ranges from 10 to 60 μg.

Except in the vicinity of cadmium-emitting industries, the cadmium in most U.S. drinking- water supplies is less than 1 μg/L (Konz and Walker 1979). However, concentrations of up to 10 μg/L have been reported in some water supplies (EPA 1981). Thus, daily cadmium intake from drinking water is about 2-20 μg, assuming that a person drinks 2 L of water per day.

Cigarettes are also an important source of cadmium exposure. The amount of cadmium that can be inhaled from smoking one cigarette is 0.1-0.2 μg (Elinder and others 1983; Friberg and others 1974). Thus, it can be estimated that someone smoking one pack per day will take in 2-4 μg of cadmium per day. Environmental tobacco smoke or passive smoking is another source of human exposure to cadmium. It has been reported that the amount of cadmium (presumably as CdO) released in the mainstream smoke (the smoke that the smoker inhales) from smoking one cigarette is about 100 ng and the cadmium released in the sidestream smoke (the smoke that originates from the smoldering end of a cigarette in between puffs) is about 720 ng (NRC 1986; DHHS 1986; EPA 1992). The cigarette-smoker is subjected to the sidestream as is anyone near the smoker while the cigarette is burning. The amount of cadmium inhaled from passive smoking depends on several factors, such as number of cigarettes smoked, ventilation, dilution factor (e.g., size of the room), and duration of exposure.

Thus, the total daily human cadmium intake in industrial countries from environmental and industrial sources is 12-84 μg/d for an adult. For a 70-kg person, that corresponds to a potential intake of 0.2-1.2 μg/kg per day. Figure 5-1 shows the typical ranges of daily cadmium intake by exposure pathway. Food products and water contribute almost all the typical daily

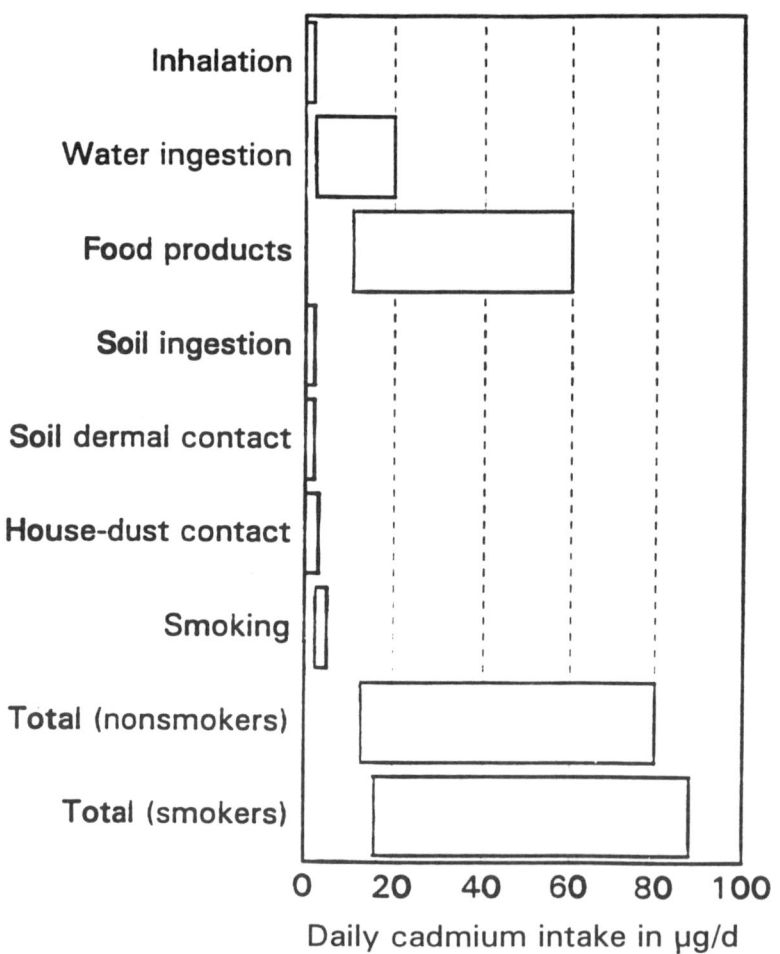

FIGURE 5-1. Typical ranges of daily cadmium intake by exposure pathway.

human exposure; inhalation contributes a very small fraction. Not all the inhaled or ingested soluble cadmium is absorbed into the body. The subcommittee used the information on water-soluble cadmium compounds as the worst case. Only 25-50% of the inhaled cadmium is absorbed by the lungs (Elinder and others 1976; Friberg and others 1986; Henderson and others 1979), and only 5% of the ingested cadmium is absorbed in the gastrointestinal tract (ATSDR 1992; Friberg and others 1974; IARC 1993; Nriagu 1980). Figure 5-2 presents a comparison of typical daily and annual inhalation intake of cadmium in urban areas with the time-integrated (total) potential inhalation doses of cadmium in the form of ZnCdS in 19 locations where it was released.

To consider both direct inhalation exposures and indirect exposures (ingestion and through skin contact) in its risk assessment, the subcommittee developed potential dose ratios that relate the direct-inhalation potential dose to the indirect potential doses associated with the same air concentration. The ratios are developed in Appendix I and summarized in Table 5-2, and were used to develop the comparisons in Figure 5-2. The exposure of a population to a cloud of ZnCdS is expressed as the product of time and concentration with units of $\mu g\text{-}h/m^3$. Because the concentration varies in time, we believe that the cumulative exposure-time product, rather than peak concentration, is the most appropriate way to express exposure. Because the population doses that result from this exposure accumulate by multiple pathways, we developed a dose-to-exposure ratio for each pathway. These ratios are shown in Table 5-2. The potential doses can be interpreted as follows. Consider the inhalation pathway, listed at the top of the table. The entry of 1 μg per $\mu g\text{-}h/m^3$ of exposure means that someone exposed for 1 h to a concentration of 1 $\mu g/m^3$ will get a cumulative dose of 1 μg. It also means that someone exposed to 0.5 $\mu g/m^3$ for 2 h, which is equivalent to an exposure of 1 $\mu g\text{-}h/m$, will receive the same dose, 1 μg. A similar approach applies to each table entry.

Appendix I provides a detailed analysis of sources of environmental and industrial exposure, transport and environmental fate, sinks (reservoirs) of cadmium, and exposure pathways.

Estimated average total daily intake from environmental and industrial exposures to cadmium from all media (soil, water, food, and air) in urban areas are greater than any daily exposures to cadmium resulting from the ZnCdS particles in the Army tests, with the exception of 1 test at Biltmore Beach, an isolated beach area in Florida. It should be noted that the cadmium intake directly from air (that is, via inhalation) contributes very little to total cadmium intake in urban and industrial areas. Although on the

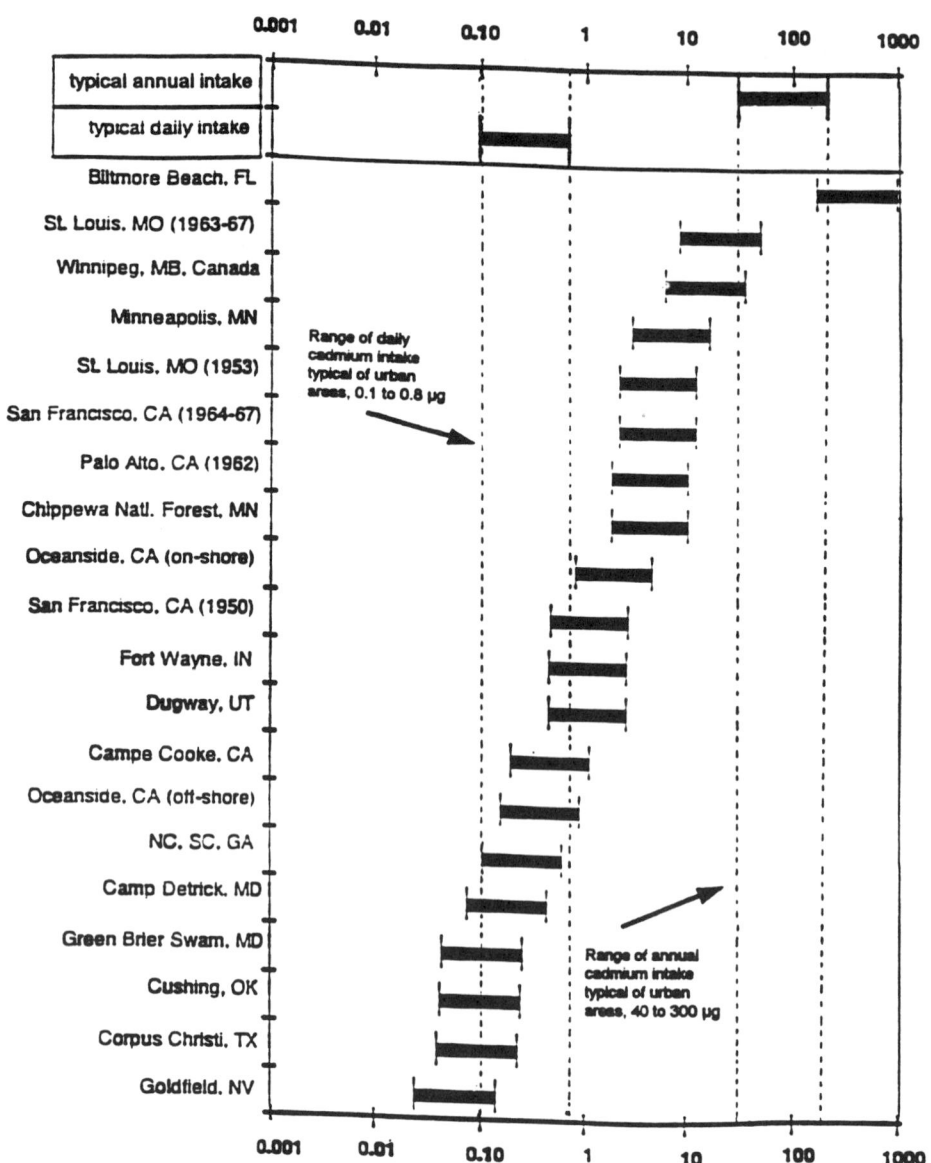

FIGURE 5-2. Comparison of typical daily and annual inhalation cadmium intake in urban areas to the time-integrated (total) potential inhalation doses to cadmium in the form of ZnCdS in locations where ZnCdS was released.

days of the Army tests, the sites of the highest monitored ZnCdS concentrations had concentrations of airborne cadmium (in the form of ZnCdS) that were above the estimated urban average daily airborne cadmium in about half the test sites, these short-term high concentrations had minimal impact on indirect cadmium exposures (from water, food, and soil).

TABLE 5-2 Summary of Potential Dose Ratios for Direct and Indirect Exposure Pathways

Exposure Pathway	Direct or Indirect Contact with Air	Potential Dose Ratio, $\mu g/((\mu g\text{-}h)/m^3)$ Cumulative Air Exposure
Inhalation	Direct	1.0
Inhalation of resuspended soil outdoors	Indirect	0.0005 to 0.002
Dermal contact with and ingestion of house dust	Indirect	2.5 to 5
Inhalation of resuspended house dust	Indirect	≈ 0.12
Deposition onto vegetation in home gardens	Indirect	2.2
Deposition onto surface drinking water supplies	Indirect	0.24 to 0.8

CONCLUSIONS

The subcommittee concludes that the four highest potential inhalation doses of ZnCdS that humans were exposed to during the Army's dispersion tests occurred in Minneapolis (44 µg), Winnipeg (93 µg), St. Louis (156 µg), and Biltmore Beach (2,500 µg). The corresponding amounts of cadmium in these doses are 6.8 µg in Minnesota, 14.5 µg in Winnipeg, 24.4 µg in St. Louis, and 390 µg in Biltmore Beach. On the basis of these estimates, the subcommittee concludes that exposure to cadmium from the dispersion tests (except for Biltmore Beach, FL, an unpopulated remote area at the time of the Army's tests) did not exceed the background exposures encountered in urban areas. It should be noted that cadmium intake directly from air (that is, via inhalation) contributes very little to total cadmium intake in urban and industrial areas. Although, on the days of the Army tests, the sites of the highest monitored ZnCdS concentrations had concentrations of airborne cadmium (in the form of ZnCdS) that were above the estimated urban average daily airborne cadmium in about half the test sites, these short-term high concentrations had minimal impact on indirect cadmium exposures (from water, food, and soil).

6

RISK CHARACTERIZATION OF EXPOSURES TO ZINC CADMIUM SULFIDE

RISK IS THE PROBABILITY of a specific outcome (such as an adverse health effect) under a particular set of conditions. Human-health risk assessment entails the evaluation of toxicological and related information on the environmental agents and the extent of human exposure to those agents. The product of the evaluation is a statement regarding the probability that populations so exposed will be harmed and to what degree.

A 1994 NRC report describes a 4-step analytic process for performing a human-health risk assessment, which involves hazard identification, dose-response assessment, exposure assessment, and risk characterization. When a substance leaves a source (such as an industrial facility), it moves through an environmental medium (such as the air), and results in an exposure (for instance, people breathe air that contains the substance). The exposure creates a dose in the exposed people (the amount of the chemical entering the body, which may be expressed in any of several ways). The magnitude, duration, and timing of the dose determine the extent to which the toxic properties of the substance are realized in exposed people (the risk).

The subcommittee reviewed toxicological and related information on ZnCdS and cadmium compounds and a vast amount of exposure data from the Army's ZnCdS dispersion tests conducted in various US and Canadian locations. Results of the subcommittee's analyses can be found in Chapters 2, "Input from the Public"; 3, "Toxicity and Related Data on Zinc Cadmium Sulfide"; 4, "Toxicity and Related Data on Selected Cadmium Compounds"; and 5, "Exposure Assessment." Greater detail on exposure data can be found in Appendix B, "A Summary of Doses and Concentrations of ZnCdS Particles from Army's Dispersions Tests."

This chapter combines information from the hazard-identification and dose-response assessments, as presented in Chapters 3 and 4, with the exposure-assessment, as presented in Chapter 5, to determine what magnitude of human exposure to ZnCdS might produce adverse health effects. The noncancer and cancer risk assessments of exposures to ZnCdS are discussed below.

RISK ASSESSMENT FOR NONCANCER HEALTH EFFECTS

Establishing health risk for exposures to agents that might be associated with noncancer health effects is based on the concept that there is a level of exposure below which no adverse health effect would be expected to occur—the so-called threshold dose. The quality of such a prediction is highest when it is based on epidemiologic or clinical studies carried out under conditions that most closely mimic the exposure situation being evaluated. When human-exposure information is not available, one must attempt to extrapolate from the best existing database, often using animal studies. One must recognize possible differences in dose, exposure route, duration, species, and chemical and physical properties of the test material. Risk assessors usually assume that humans and the most sensitive animal species are equally sensitive to the test chemical unless evidence indicates otherwise. The most scientifically defensible assessments are based on the best available information from a combination of human and experimental-animal studies.

Accuracy and precision in estimating noncancer human health risk are largely determined by the quality and quantity of the toxicologic data available for analysis. A wide variety of noncarcinogenic responses can be potentially produced by exposure to chemical substances. The responses can be minor and temporary or severe and permanent. The nature and severity of a health effect can depend on several factors, such as the amount of the chemical to which a person is exposed and the potency or toxicity of the chemical itself. Other factors peculiar to the exposed person can influence the effect, including age, sex, health status, lifestyle, and diet. The health effects depend on the route by which the chemical enters the body, the changes in the chemical as it moves through the body, and the specific target organ that is the most susceptible.

Noncancer risk assessment normally incorporates a number of "safety" or "uncertainty" factors, which are applied when there is a need to accommodate human response variability, including response in sensitive subgroups; to predict human response from animal data; to extrapolate from subchronic to chronic exposure; to predict a no-observed-adverse-effect level (NOAEL) from a lowest-observed-adverse-effect level (LOAEL); or to use a database that is considered incomplete. Factors ranging between 1 and 10 are most often used for each of the sources of uncertainties listed above, depending on the nature and sensitivity of the adverse effect in question. As the data become more uncertain or unreliable, higher safety factors are applied. Any toxic response might be used for establishing the NOAEL or LOAEL as long as it is the most-sensitive toxic effect and is considered likely to occur in humans. The exposure limit for noncarcinogenic health effects is usually derived from the animal-test data that leads to a NOAEL. This standard approach seems appropriate and was used by the subcommittee for assessing the potential noncancer health risk associated with exposure to ZnCdS.

The toxicity database on ZnCdS is not adequate to be used in risk estimates for human exposure. Therefore, toxicity information on some other related chemical can be used to estimate the risk associated with ZnCdS exposures. The subcommittee considered three possible scenarios to estimate the noncancer risk associated with exposure to ZnCdS from the Army's dispersion tests.

In the worst-case scenario, ZnCdS would have the toxic properties of soluble cadmium compounds. However, the physical and chemical properties of ZnCdS are known. It is insoluble in water and lipids and only poorly soluble in strong acids. If it is assumed that in vivo ZnCdS becomes as soluble as any cadmium compound and might release its cadmium ions to react freely with biologic targets, it would be appropriate to estimate the toxicity of ZnCdS from the toxicity of soluble cadmium compounds in general. It is extremely doubtful whether such an assumption is warranted. Our general understanding of metal toxicology gives this worst-case scenario little, if any, plausibility.

In a second scenario, ZnCdS might actually be a biologically inert particle, not more toxic than all the other respirable particles present in our daily environment. This scenario is based on physicochemical properties of ZnCdS (only poorly soluble in strong acids), which results in a lack of acute oral or dermal toxicity. However, the lack of a comprehensive evaluation of the toxicity of ZnCdS, particularly long-term low-level effects, and the fact that ZnCdS might be degraded to some extent, albeit only very slowly, preclude endorsement of such an assumption.

In the third scenario, ZnCdS would have toxic properties similar to those of cadmium sulfide, CdS, an insoluble cadmium compound. ZnCdS has a crystalline structure similar to that of CdS; the only difference is that in ZnCdS, zinc replaces 80% of the cadmium in the lattice. CdS is insoluble in water and lipids and slightly soluble in strong acids. Experimental data on toxicokinetics and toxicity of CdS are available. They clearly show that the cadmium in CdS is much less bioavailable than the cadmium in soluble compounds. Furthermore, it is plausible that the fusion of CdS with zinc sulfide, ZnS, at 900°C will produce a crystalline lattice structure from which cadmium is even less bioavailable than CdS. Until experimental data to the contrary become available, it may reasonably be assumed that the fusion of CdS with ZnS essentially preserves the physicochemical and hence toxic properties of CdS. Therefore, the subcommittee chose to base its assessment of the potential toxicity of ZnCdS for noncancer health effects on the toxicity of CdS.

Results of studies conducted by Oberdörster (1990) and Oberdörster and Cox (1990) indicate that CdS inhaled into the lung is not solubilized and absorbed into the blood for distribution to other organs. Therefore, the subcommittee concluded that evidence suggests that sporadic exposures of the public to small amounts of ZnCdS should not result in any systemic toxicity (toxicity to the kidney, bone, immune system, or reproductive and developmental system).

The only organ system considered to be potentially at risk because of exposure to particles of ZnCdS was the respiratory tract; the particles of ZnCdS were respirable and would be expected to deposit deep in the lungs. Results of studies of Oberdörster and others (1985) and of Glaser and others (1986) indicate that CdS has little toxicity in the respiratory tract. The subcommittee used the studies of Glaser and others (1986) as the basis for their risk assessment because the cumulative exposures were higher in that study. In the Glaser and others (1986) studies, rats exposed to CdS at 39,600 mg-min/m^3 had only a mild pulmonary response. Using those data and dividing by an uncertainty factor of 10 to extrapolate from a LOAEL to a NOAEL, by a factor of 10 to extrapolate from animal to human exposures, and by another factor of 10 to account for sensitive populations, one would not expect adverse health effects, even in sensitive populations, from exposure to cadmium sulfide in an insoluble form, such as ZnCdS, at 39.6 mg-min/m^3 or 39,600 µg-min/m^3.

This level of 39,600 µg-min/m^3 CdS (inhaled dose of 594 µg) or 30,900 µg-min/m^3 cadmium (39,600 × 0.78 = 30,900) or 513 µg of inhaled cadmium dose (30,900 µg-min/m^3 × 0.0166 m^3/min = 513 µg) is considered to be a "safe" level of exposure and was compared with the maximal inhaled doses at the test sites. The four highest estimated inhaled cadmium doses were 6.8 µg (Minneapolis), 24.4 µg (St. Louis), 14.5 µg (Winnipeg), and 390 µg (Biltmore Beach, FL). Inhaled cadmium doses were derived by multiplying the maximal exposures to ZnCdS (expressed as µg-min/m^3 and shown in Table B-1 of Appendix B) by 0.0166 m^3/min (the volume of air inhaled by an active person in 1 min). The product is then multiplied by 0.156 (the proportion of cadmium in ZnCdS). For example, the maximal ZnCdS exposure in Biltmore Beach, FL was 150,000 µg-min/m^3. Multiplying this by 0.0166 m^3/min yields an inhaled dose of ZnCdS of about 2,500 µg. Multiplying that by 0.156 yields an inhaled cadmium dose of about 390 µg. Thus, inhaled cadmium doses from

ZnCdS tests did not exceed the "safe" level (513 µg) at any test site. On the basis of the above information, the subcommittee concludes that the exposures to ZnCdS from the Army's tests (for which data are available) did not increase people's exposure to cadmium over background levels by enough to have caused noncancer health effects in exposed persons.

RISK ASSESSMENT FOR CANCER

The cancer risk associated with a particular chemical is estimated by multiplying the total exposure to it by its cancer-potency factor.[1] The subcommittee based its cancer risk estimates on the cadmium content of ZnCdS because no studies on the carcinogenicity of ZnCdS have been conducted and cadmium is the most-toxic component of ZnCdS. As stated previously, this approach will therefore overestimate the risk. Assuming that 100% of the cadmium is bioavailable, estimates of lung-cancer risk were calculated for all test sites for which air concentrations are available based on the maximal recorded cumulative exposures of ZnCdS in the air and the carcinogenic potency derived from a study of workers exposed to cadmium by inhalation and oral ingestion of dust containing cadmium compounds. Table 6-1 summarizes the assumptions made in calculating the cancer risk estimates.

The cancer risk estimates are based on the total cumulative exposure to cadmium in ZnCdS. As will be shown later, the cancer-potency risk factor was derived from the occupational exposure of workers to cadmium. That provided an estimate of the human potency factor expressed as lung-cancer risk per milligram of cadmium. Because workers were exposed to cadmium compounds at higher air concentrations for periods much longer than those of any people exposed from the ZnCdS releases, this cancer-potency risk factor might overestimate the risk for much shorter exposures to ZnCdS.

---

[1]Cancer-potency factor is a value used by regulatory agencies to describe the inherent potency of carcinogens. The factor is derived with dose-response models. Upper limits on cancer-potency factors, also called q_1^*, are used to estimate an upper limit on the likelihood that lifetime exposure to a particular chemical could lead to excess cancer deaths.

Risk Characterization

TABLE 6-1 ZnCdS Cancer Risk Estimates: Major Assumptions Made and Their Effects on Lung-Cancer Risk Estimates

Assumption or Condition	Impact
100% of cadmium in ZnCdS was bioavailable	If only a fraction of cadmium is bioavailable, all risks are overestimated
Cancer potency of cadmium based on exposures of adult workers; cancer-risk estimates multiplied by 10 to account for possible higher sensitivity in some subpopulations, such as the young or elderly	Overestimates cancer risks by a factor of 10 for most exposed people
Short-term exposure results in a cancer risk proportional to the duration of exposure (total dose)	Could overestimate cancer risk by an unknown margin
Sampling time was adequate to cover the total airborne exposure for a test release	If sampling time were inadequate, dose could have been underestimated
The best estimate of the cumulative exposure at the maximal-exposure location is used; the best estimate is likely to be within a factor of 2 of the maximal cumulative exposure at any fixed location	Increases the estimate of risk because of the premise that an individual remains at the maximual-exposure location for an extended period
Conservative assumptions regarding lung-cancer potency and dose of ZnCdS were used, with maximal recorded cumulative exposures at a test site, so upper bounds of lung-cancer risk were estimated	It is likely that the lung-cancer risk is overestmated for most, if not all, people; the cancer risk might approach zero

Traditionally, the cancer-potency factor and exposure are both expressed on the basis of average daily exposure over a lifetime. That does not imply here that a lifetime exposure is required to produce cancer. As used here, it is only a dosimetric convention that does not affect the estimation of risk. The average daily exposure is directly proportional to the cumulative exposure on which the cancer risk estimate is based.

The cancer risk estimates were calculated first on the basis of a lifetime of exposure and then reduced to the proportion of a lifetime during which people were actually exposed in the test areas. This is equivalent to using the daily exposure averaged over a lifetime. The procedure of multiplying the risk for a lifetime exposure by the fraction of actual exposure is based on calculations with the multistage model of carcinogenesis (Kodell and others 1987) and the Moolgavkar-Knudson-Venzon 2-stage clonal-expansion model (Chen and others 1988), which show that it is not likely to underestimate the risk associated with short-term exposure by more than a factor of 10. That was verified by a comparison of tumor risks related to lifetime and less-than-lifetime exposures reported for animal studies (Gaylor 1988). Cancer risk estimates were multiplied by 10 to account for exposure during possibly sensitive ages, such as childhood. It is unlikely that the procedures used underestimated the lung-cancer risks associated with the reported exposure to cadmium (in ZnCdS) from the Army tests.

LUNG-CANCER POTENCY OF CADMIUM

Thun and others (1991) report excess lung-cancer risk in workers exposed to various amounts of cadmium by inhalation, namely, differential risks from a standard population of -1.77%, +2.87%, and +14.23% for median exposures of 767; 3,315; and 11,507 μg-yr/m^3, respectively. They offer several possible reasons for the reduction in lung cancer for the low-exposure group, including the possibility of a select healthy-worker effect. Regardless, the relationship between excess cancer rates and exposures to cadmium compounds in these workers provides an estimate of lung-cancer potency of inhaled cadmium compounds. A simple linear regression provides a potency estimate of 14.6×10^{-6} per μg-yr/m^3. A worker inhaling air at 10 m^3/d, 5 d/wk, for 48 wk/yr inhales $10 \times 5 \times 48 = 2,400$ m^3/yr. Thus, the lung-cancer potency estimate is $(14.6 \times 10^{-6})/ 2,400 = 6.1 \times 10^{-6}$ per milligram of cadmium.

An exposure of a 70-kg person to 1 mg over 75 yr amounts to a lifetime average daily exposure of $1/(70 \times 365 \times 75) = 0.52 \times 10^{-6}$ mg/kg-day. The cancer-potency risk factor and exposure are both expressed on a body-weight basis. Here, this has no effect on the estimate of risk. The cancer risk estimate is based on the cumulative exposure of humans. Thus, there is no need for a cross-species dose-scaling factor based on, for example, lung surface area. The lung-cancer potency estimate expressed in terms of lifetime average daily exposure is $(6.1 \times 10^{-6})/(0.52 \times 10^{-6}) =$

11.7 per milligram of cadmium per kilogram body weight per day. The estimate of cancer potency based on human data is almost twice the upper limit for potency of 6.3 per milligram per kilogram body weight per day of cadmium based on inhalation studies in rodents (Integrated Risk Information System, Office of Health and Environmental Assessment, EPA, Washington, D.C.).

The studies in workers exposed to cadmium do not indicate which particular compounds are responsible for the carcinogenic response. The potency value for lung cancer might reflect "bioavailable" cadmium, but it might also reflect cancer caused by the specific cadmium compounds present. It is reassuring that the cancer potency estimates for cadmium based on animal or human data are within a factor of 2 of each other. The National Research Council's report *Science and Judgment* (NRC 1994) states that "the carcinogenic risk associated with specific cadmium compounds could be overestimated or underestimated, because bioavailability has not been included in the risk assessment." However, the Occupational Safety and Health Administration came to the conclusion (OSHA 1992): "Record evidence and expert opinion [given by G. Oberdörster and U. Heinrich] led the Agency to conclude that CdS should be considered an occupational carcinogen and assigned the same PEL [permissible exposure limit] as that established for other Cd compounds."

In estimating the potential lung-cancer risk associated with exposure to ZnCdS, the subcommittee followed the procedure adopted by OSHA and assumed equal bioavailability of cadmium from all cadmium compounds. On the basis of the subcommittee's expert judgment and the data presented in Chapters 3 and 4, it is highly unlikely that this is the case for ZnCdS.

LUNG-CANCER RISK ESTIMATES

Releases of ZnCdS at all locations are shown in Table 5-1 and Appendix B, Table B-1. Maximal cumulative exposures reported at a site were expressed, in terms of ZnCdS, as microgram-minutes per cubic meter. Multiplying this by 0.0166 m^3/min yields an inhaled dose of ZnCdS. Multiplying that by 0.156 yields an inhaled cadmium dose. The lifetime average daily exposure was calculated by dividing by 27,375 days (365 × 75) and by 70 kg to convert to a body-weight basis. The average daily exposure expressed as milligrams of Cd per kilogram per day was multiplied by the human lung-cancer potency factor, 11.7, to obtain an estimate of the lung-cancer risk. The estimates were multiplied by 10 to account for possible increased sensitivity of children (Table 6-2).

For example, the maximum potential exposure reported for Biltmore Beach, FL was 150,000 μg-min/m^3 of ZnCdS (Appendix B, Table B-1). Using an inhalation rate of 1m^3/h, an active person inhales 1/60th of a cubic meter of air per minute, i.e., 1/60 = 0.0166 m^3 per minute. The maximum potential exposure was 0.0166 × 150,000 ≈ 2,500 μg of ZnCdS. Because 15.6% of ZnCdS is cadmium, the maximum potential exposure to cadmium is 0.156 × 2,500 = 390 μg, as reported in Table 6-2. Averaged over a lifetime of 75 years × 365 days per year = 27,375 days, the average daily exposure is 390 divided by 27,375 = 0.0142 μg of cadmium per day. For a body weight of 70 kg (148 pounds), the average daily exposure per kg of body weight is 0.0142 divided by 70 = 0.0002 μg/kg of cadmium. The human lung-cancer potency factor (probability of lung cancer per mg/kg cadmium per day average lifetime exposure) calculated earlier in this chapter is 11.7 per mg/kg of cadmium per day over a lifetime. Since one μg is 1/1,000th of a mg, the potency factor is 11.7 divided by 1,000 = 0.0117 per μg/kg of cadmium per day. The lifetime risk (probability of lung cancer) is estimated to be less than the potency factor times the average daily lifetime exposure: 0.0117 × 0.000204 = 0.0000024 = 2.4 × 10^{-6} (2.4 per million). The estimates were multiplied by 10 to account for possible increased sensitivity of children, 10 × 2.4 × 10^{-6} = 24 × 10^{-6} (24 per million) as reported in Table 6-2. For this particular remote beach area, it is unlikely that anyone was exposed at this level.

The highest cancer risk estimates are at Biltmore Beach in the vicinity of Panama City, FL. Air samples there were collected on a remote beach over an area several yards wide in the center of the plume. It is unlikely that many people, if any, were exposed to such doses.

TABLE 6-2 Upper-Bound Estimates of Cumulative Airborne Cadmium Exposure and Lung-Cancer Risk Associated with Exposures of Potentially Sensitive Populations (Such as Children) at Locations with Maximal Recorded Exposure[a]

Exposure Site	Estimated Maximal Exposure per Person, μg	Estimated Maximal Lifetime Risk[c]
Biltmore Beach, FL[b]	390	24.0×10^{-6}
St. Louis, MO (1963-1965)	19.2	1.2×10^{-6}
Winnipeg, MB, Canada	14.5	0.9×10^{-6}
Minneapolis, MN	6.8	0.4×10^{-6}
St. Louis, MO (1953)	5.2	0.3×10^{-6}
San Francisco, CA (1964-1967)	4.9	0.3×10^{-6}
Palo Alto, CA (1962)	4.4	0.3×10^{-6}
Chippewa National Forest, MN	4.2	0.3×10^{-6}
Oceanside, CA (on shore)	1.8	0.1×10^{-6}
San Francisco, CA (1950)	1.1	0.07×10^{-6}
Ft. Wayne, IN	1.1	0.07×10^{-6}
Dugway, UT	1.1	0.07×10^{-6}
Clinton School, Minneapolis	0.5	0.05×10^{-6}
Camp Cooke, CA	0.4	0.03×10^{-6}
Oceanside, CA (off shore)	0.4	0.03×10^{-6}
NC, SC, GA	0.3	0.02×10^{-6}
Camp Detrick, MD	0.2	0.01×10^{-6}
Greenbrier Swamp, MD	0.1	0.01×10^{-6}
Cushing, OK	0.1	$<0.01 \times 10^{-6}$
Corpus Christi, TX	<0.1	$<0.01 \times 10^{-6}$
Goldfield, NV	<0.1	$<0.01 \times 10^{-6}$
Colfax, WA	<0.1	$<0.01 \times 10^{-6}$
Stanford Univ., CA	<0.1	$<0.01 \times 10^{-6}$
Palo Alto, CA (1950)	<0.1	$<0.01 \times 10^{-6}$

[a]Almost all the people had exposures (and risks) lower by a factor of 2-10. Calculations assumed 100% bioavailability of cadmium from ZnCdS.

[b]It is unlikely that many people, if any, were exposed to such doses because it was a remote, unpopulated beach area.

[c]Twenty-four additional cancers in 1 million persons.

The highest estimates of lung cancer risk for a populated area were for St. Louis, MO. The Army conducted tests in St. Louis in 1953 and again during 1963-1965. It is possible that some individuals were in the maximally exposed areas during both tests. Their total maximum exposure could have been $(19.2 + 5.2) = 24.4$ µg of cadmium with an estimated lifetime lung cancer risk of less than $(1.2 + 0.3) \times 10^{-6} = 1.5 \times 10^{-6}$.

Examination of records and isopleth maps of the levels of ZnCdS measured in the air indicate that the maximum exposures occurred over a very small fraction of the test areas. For rooftop and street-level releases of ZnCdS, the vast majority of the population were exposed at less than 1/10th the maximum recorded values. Hence, the estimated lung-cancer risks for the vast majority of these populations are less than 1/10th that of the maximally exposed individuals as reported in Table 6-2. For airplane releases of ZnCdS, the vast majority of the population were exposed to less than one-half the maximum recorded values. Hence, their estimated lung-cancer risks are half (or less than half) that reported in Table 6-2 for the maximally exposed individuals.

UNCERTAINTY OF LUNG-CANCER RISK ESTIMATES

The major assumptions made and their effects on cancer risk estimates are listed in Table 6-1. It was assumed that all the cadmium in the ZnCdS particles was bioavailable. If only a fraction, such as 50%, of the cadmium were bioavailable, the risk estimates would be half of those calculated. The lung-cancer potency was estimated from results in workers exposed to cadmium compounds at higher air concentrations for periods much longer than those of any people exposed in the ZnCdS releases. It is unknown whether the cancer potency (risk) is lower for shorter periods. The values chosen for the breathing rate of active people (1 m^3/h) and life span (75 yr) have little influence on cancer risk estimates.

Undoubtedly, some people were exposed to cadmium concentrations somewhat higher than those recorded at any air sampler location. At Minneapolis and Ft. Wayne, where cadmium air concentrations were estimated from isopleths, it is unlikely that exposure could have exceeded the reported values by a factor of 10. In fact, lung cancer risk estimates are unlikely to be low by a factor of 10 due to underestimates of air concentrations for the maximum exposed individuals.

OTHER ROUTES OF EXPOSURE

The cancer risk estimates in this section were based on direct exposure to airborne ZnCdS inhaled during the test releases. Some ZnCdS from the test releases would also settle on the ground and result in exposure via inhalation of recirculated dust particles. The workers on whom the cancer potency estimate was based were also exposed to recirculated dust particles. Hence, the contribution of this route of exposure, if any, to lung cancer was included in the potency estimate.

CONCLUSIONS

In the absence of adequate toxicity data on ZnCdS, the subcommittee considered it to be most appropriate to base its assessment of the potential toxicity of ZnCdS for noncancer health effects on the toxicity of CdS, an insoluble cadmium compound. Results of several toxicity studies conducted with CdS indicate that CdS inhaled into the lung is not solubilized and absorbed into the blood for distribution to other organs. Therefore, the subcommittee concluded that sporadic exposures of the public to small amounts of ZnCdS should not result in any noncancer systemic toxicity (toxicity to the kidneys, bone, immune system, or reproductive and developmental system). The only organ considered to be potentially at risk because of exposure to particles of ZnCdS was the respiratory tract; the particles of ZnCdS were respirable and would be expected to deposit deep in the lungs. Interaction of lung tissue with inhaled particulate matter might create circumstances that would assist in metal mobilization, and the release rate could be slow. Results of several toxicity studies indicate that CdS has little toxicity in the respiratory tract. Rats exposed to CdS at 39,600 mg-min/m^3 had only a mild pulmonary response. Using that result and dividing by 10 to extrapolate from a LOAEL to a NOAEL, by 10 to extrapolate from animal to human exposures, and by 10 to account for sensitive populations, one would not expect adverse health effects, even in sensitive populations, from exposure to CdS in an insoluble form, such as ZnCdS, at 39.6 mg-min/m^3 or 39,600 μg-min/m^3 or inhaled cadmium dose of 513 µg.

The inhalation dose of 513 µg was considered to be a safe level of exposure and was compared with the maximal cadmium exposures at the test sites. The four highest cadmium doses that could have been inhaled were estimated to be 6.8 μg (Minneapolis), 24.4 μg (St. Louis), 14.5 μg (Winnipeg), and 390 μg (Biltmore Beach, FL). The exposures to cadmium from ZnCdS did not exceed the level considered to be safe at any test site. On the basis of the above information, the subcommittee concludes that the exposures to ZnCdS from the Army's tests (for which data are available) should not have caused noncancer health effects in exposed persons.

Cancer risk estimates are based on the cadmium content of ZnCdS because no studies on the carcinogenicity of ZnCdS have been conducted and cadmium is the most-toxic component. This approach will overestimate the risk. Upper-bound estimates of lung-cancer risk based on 100% bioavailability of the inhaled cadmium from ZnCdS particles at the maximal recorded exposure are summarized in Table 6-2. For each city or town, the location with the highest average reported air concentration of Cd was used to estimate cancer risk. The risk estimates were multiplied by a factor of 10 to account for exposure to potentially sensitive subpopulations, such as children. The subcommittee believes it is unlikely that lung-cancer risk estimates have been underestimated. The potential exposures of people with the highest ZnCdS exposures from the Army tests in St. Louis were equivalent to the ambient airborne cadmium exposures of people living in typical urban areas for 1-8 mo. For the vast majority of people exposed to ZnCdS from the Army tests in St. Louis, the exposures were equivalent to the ambient airborne exposures of people living in typical urban areas for 1-3 wk. The estimated upper-bound lung-cancer risks are small even for people with the larger exposures.

On the basis of a description of the tests and, in some cases, isopleths of exposure, the vast majority of people were exposed at concentrations that were less than the highest by a factor of 2-10 or even more. Accordingly, their cancer risks would be lower by a factor of 2-10 or more. It is unlikely that any people have experienced total exposures to cadmium much above the exposures that would have been experienced by a person who was at the maximally exposed location for each of the ZnCdS tests in a particular locale. It is unlikely that exposure estimates have been underestimated for the vast majority of the exposed population. Because of the extremely low concentrations of ZnCdS (or cadmium from ZnCdS) in the air and the short duration of exposure, the lung cancer risk, if any, is very low. Hence, it is unlikely that anyone in the test areas developed lung cancer owing to direct exposure to cadmium from airborne releases of ZnCdS.

7

Scientific Feasibility of Epidemiologic Study

IN THIS CHAPTER, the subcommittee addresses the utility and feasibility of conducting epidemiologic studies to assess whether adverse human health effects could be associated with exposure to ZnCdS as a result of the Army's dispersion tests. This assessment was performed by the subcommittee in response to citizens who requested that health studies be conducted in their communities.

The chapter has 3 sections. The first section introduces the field of epidemiology, providing an overview of what it can do and its limitations, its underlying assumptions, and the types of research designs that it uses. The second discusses key methodologic issues that need to be considered in deciding whether an epidemiologic study is feasible and likely to produce scientifically valid information about a particular exposure and health outcome. The last section discusses the types of epidemiologic studies that were considered by the subcommittee in light of available information, including that provided by members of the public at public hearings.

NATURE OF EPIDEMIOLOGIC INVESTIGATIONS

Epidemiology can be broadly defined as "the study of the distribution and determinants of disease frequency" in human populations (MacMahon and Pugh 1970). Implicit in that definition is the assumption that disease does not occur randomly and that systematic epidemiologic methods can be useful for identifying risk factors for various diseases. Epidemiology can be used to assess empirical associations between an exposure and disease, especially when the increase in risk of a disease attributed to an exposure is large, when the exposure can be measured with objective criteria, and when exposed and unexposed people can be clearly distinguished. Epidemiologic studies are less informative when the difference in risk between exposed and unexposed people is small, when exposures are poorly defined or measured, when exposed and unexposed people are not readily distinguished, or when confounding factors exert a much greater effect on the incidence of the disease than does the exposure of interest. An individual's risk of disease cannot be determined through epidemiologic study; rather, such study determines average risks for populations or at least for representative samples.

In general, epidemiologic studies are designed to study risk factors for a specific disease or the possible health effects associated with a specific exposure. In recent years, environmental epidemiology has developed as a subspecialty of epidemiology. It focuses on the role of physical, biologic, chemical, and psychosocial factors—factors not necessarily under the individual's control—and human health (NRC 1991a; Rothman 1993). Environmental epidemiologists are increasingly asked about past exposures and their relation to present health problems.

Various research designs are available to epidemiologists, depending on whether a study begins with a particular disease outcome or with an exposure. Retrospective or case-control studies categorize people on the basis of whether they have the disease of interest (cases) or are free of the disease (controls) at the time of study. Controls should be similar to cases except for the presence of the disease and need to be carefully selected for study; the two groups are compared with respect to the proportions that have had a particular exposure. Most information is collected retrospectively for case-control studies. Prospective or cohort studies identify populations or samples for study with respect to an exposure and follows them over time to identify the onset of disease. Information collected prospectively is considered by some to be more reliable than information collected retrospectively. Regardless of design, epidemiologic studies need to conform to the scientific principles of research methodology if their findings are to be considered scientifically valid and useful.

KEY METHODOLOGIC ISSUES

Three key methodologic issues would have to be considered together in designing a scientifically valid epidemiologic study that would be capable of answering the concerns of people who were exposed to ZnCdS: (1) Is a well-defined population comprising clearly defined exposed and unexposed individuals available for study? (2) Is the exposure capable of causing the suspected health outcomes? (3) Is the other information needed for a well-designed epidemiologic study available? These issues are discussed below as they relate to a possible study of health effects of exposure to ZnCdS.

1. *Is a well-defined population comprising clearly defined exposed and unexposed individuals available for study?*

Accurate measurement of an exposure is a key aspect of any epidemiologic study. In environmental epidemiology, the crudeness with which an exposure is measured often renders the results inconclusive. A somewhat different set of circumstances surrounds the Army's testing with ZnCdS. Extensive data on ZnCdS dispersion patterns are available for several geographic areas—including Minneapolis, Fort Wayne, and Corpus Christi—because the Army was interested in measuring the dispersion of the compound and monitored it closely starting soon after its release.

However, there are two major limitations to the exposure data. First, data are not available regarding the actual exposures or doses that people (as opposed to geographic areas) received as a result of the Army's tests. Exposure and dose are important concepts for environmental epidemiology, as discussed by the NRC (1989, 1991b). "Exposure" refers to the concentration of (and duration of exposure to) an agent that a person receives from the environment, whereas "dose" refers to the amount deposited in or absorbed by the body (Hatch and Thomas 1993). The subcommittee considered whether geographic residence could be used to estimate individual exposure. This approach was considered a crude means of measuring exposure, because of the number of assumptions that underlie it. For example, it assumes that all people who lived in a geographic area were there at the time of the exposure and that all people received the same exposure regardless of their exact location (for instance, inside or outside). Furthermore, ZnCdS exposure occurred over 35 years ago, so it is likely that many people could not be found or would not recall where they were at the time of the Army's testing.

A second limitation of the exposure data involves the inability to distinguish between cadmium from the Army's tests and cadmium derived from other sources. If the exposure of interest for determining health effects is exposure to cadmium, the exposure to cadmium from ZnCdS would be very low and indistinguishable from background environmental and industrial sources of cadmium (see Chapter 5). Cadmium is naturally present in soil and groundwater and has wide industrial uses, for example, in paints, plastics, and nickel-cadmium batteries. In addition to the difficulty in knowing how much cadmium might be available from ZnCdS (see Chapter 5), measurement of the cadmium exposure that would have resulted from ZnCdS would yield a highly imprecise estimate of total exposure to cadmium. If it is assumed that exposure to ZnCdS is equivalent to exposure to cadmium, almost all persons received a total cadmium exposure from the Army tests of less than 1 µg and 24.4 µg is the highest estimated total potential dose received by any person in a populated area. The average daily intake of cadmium from ambient air ranges from 0.1 to 0.8 µg in urban areas. The average daily intake of cadmium from all sources (air, water, food, and so on) ranges from 12 to 84 µg in industrial areas. Therefore, most people received a total cadmium exposure from the Army tests that was no greater than that associated with living in a typical urban area for less than a month. Such a small difference between cadmium from the

Army's ZnCdS tests and that from background sources makes it very unlikely to detect health effects in an epidemiologic study.

Measurement of internal dose is needed for analyzing potential dose-response relationships, although often this information is not available in community-wide situations. When a dose-response relationship can be observed, researchers are more likely to attribute the observed health effects to the exposure. To assess dose-response relationships, people need to be categorized as to whether they have been exposed and, if so, by the amount (dose) of the substance that was deposited or absorbed. Dose is more difficult to assess in environmental studies than in occupational studies in which workers are followed and monitored directly. No data are available for assessing individual exposure to ZnCdS. Not all exposed persons received the same dose, nor will all exposed persons manifest adverse health effects. Several characteristics will determine whether exposed people exhibit disease, such as age, genetic predisposition, immune function, and health status. In addition to dose, the duration and timing of the exposure might affect the magnitude and severity of human health effects, as discussed below.

In summary, the subcommittee recognizes that people could be classified with respect to ZnCdS exposure on the basis of residence at the time of the Army's tests. However, this approach would yield only a crude approximation of exposure and would be subject to bias stemming from difficulties in finding residents of exposed areas and from misclassification of people with respect to exposure status. Accurate measurement of dose for individuals is not possible, given the very small contribution of ZnCdS to the total exposure to cadmium.

2. *Is the exposure capable of causing the suspected health outcomes?*

As discussed in previous chapters, few, if any, human-health data are available on ZnCdS exposure from previous studies. In considering the feasibility of conducting an epidemiologic study, the subcommittee evaluated the contribution of cadmium from ZnCdS in a worst-case scenario. Much of the previous work on cadmium toxicity has focused on occupational groups in which exposures were very high and continued for long periods. Occupational exposure to cadmium is not the same as exposure to cadmium in the form of ZnCdS. That is an important distinction because it appears that when cadmium is ingested in the form of ZnCdS, it is not readily soluble in the body (Leighton 1955; Weast and others 1986).

A review of the scientific literature suggests that only a few health outcomes might be linked to cadmium exposure: lung cancer, dysfunction of the proximal renal tubules, and possibly infertility and low birth weight. The spectrum of health effects reported in the subcommittee's public hearings is much wider—infertility, mental retardation, autoimmune disease, melanoma, lymphoma, leukemia, chronic lung disease, breast cancer, joint pain, skin problems, and many others.

The subcommittee recognizes several serious limitations of research on health effects of ZnCdS. The first is the lack of previous human studies on this compound, although cadmium has been studied extensively and is known to be associated with health risks. Second, there is a lack of health information systematically collected before, during, and after the periods of exposure in the affected communities; it is essential to have data on the temporal relationship of exposure to health outcomes to establish a cause-effect relationship (Hill 1965). Third, although health outcomes can be determined in individuals who identify themselves to investigators, it would be difficult to link specific health outcomes to the ZnCdS exposures. In general, it is possible to confirm the presence of disease in individuals. It is not possible to attribute illnesses to ZnCdS exposures that occurred over 35 years ago. The long latency period between exposure and disease increases the difficulty in collecting the necessary information and makes interpretation of results in establishing causality more difficult. Residents who voluntarily came forward in public meetings might not constitute a representative sample for an epidemiologic study.

3. *Is the other information needed for a well-designed epidemiologic study available?*

A well-designed epidemiologic study requires information on attrition, statistical power, and confounders. It can be hard to find people, especially if they have moved or changed names and especially in the absence of personal identifiers, such as social-security numbers. The absence of that information in communities exposed to ZnCdS could complicate follow-up.

The subcommittee considered the issue of statistical power in assessing the feasibility of an epidemiologic study of cancer or dysfunction of the proximal renal tubules in relation to ZnCdS. Statistical power is the ability to detect a true difference in health status, given the size of a population, the prevalence of exposure, and the expected size of the effect. We specifically chose those 2 health outcomes on the basis of scientific evidence and public concern. Our maximal estimates of increased risk of cancer attributed to ZnCdS for most communities is approximately 1 additional cancer case per million people exposed (see Chapter 6). As an example, assume that even our conservative estimate of 1 extra cancer death per million people exposed is too low by a factor of 100. Could we carry out an epidemiologic study if in fact ZnCdS exposure causes 1 extra cancer death among 10,000 exposed people? Our power calculations indicate that to detect an effect of that magnitude, we would need to study more than 62 million people, half of whom are in a high-exposure category. Similarly, the subcommittee performed power calculations for lung cancer and proximal renal tubular dysfunction, which are reported to affect about 2% of the general U.S. population (American Cancer Society 1996; OSHA 1992). If we assume that exposure to ZnCdS results in 1 extra case of lung cancer or renal disease per 10,000 people, we would need to study more than 30 million people, half of whom are in a high-exposure category. That figure is several times larger than the present combined populations of Minneapolis, Fort Wayne, and Corpus Christi, where some estimates of population exposure are available.

Assessment of confounding is a major consideration for any epidemiologic study. Information on known or potential confounders—factors that are associated with both the study exposure and the outcome—needs to be collected. Potential confounders of particular relevance to any ZnCdS studies include industrial exposures in Fort Wayne, refinery exposures in Corpus Christi, and indoor air pollution in Minneapolis. Without proper control in either the design or analytic phase of research, confounders can distort the true relation between an exposure and an outcome (Kleinbaum 1994). Simply stated, uncontrolled confounders can overestimate or underestimate the magnitude of health risk associated with an exposure. For example, very high cadmium exposures have been linked to lung cancer (IARC 1993; Thun and others 1991) and, in 1 study, diminished infant size (Laudanski and others 1991). Both those outcomes also are adversely affected by cigarette smoke, which contains cadmium, as well as nicotine, tars, carbon monoxide, and other toxic substances.

After weighing various methodologic considerations, the subcommittee concluded that it is not possible to gather supporting health information, including that pertaining to possible confounders, and that an extremely large sample would be required to detect any statistically significant differences between exposed and unexposed groups. Those limitations, coupled with difficulties in identifying exposed and unexposed populations other than by crude estimates based on residence, argue against an epidemiologic study.

TYPES OF EPIDEMIOLOGIC STUDIES

The subcommittee considered several epidemiologic approaches in determining the feasibility of conducting an epidemiologic study of the health effects from ZnCdS exposure. We considered various study designs in light of the information presented by members of the public attending the public hearings. Possible study designs are discussed separately below and in relation to the three key research issues addressed above.

The subcommittee considered strategies for the retrospective follow-up of populations exposed to the Army's dispersion tests, despite the absence of individual-exposure data. Several barriers to a retrospective follow-up study were identified. Mortality or cause-of-death data will be available for communities only if death certificates were filed with particular states' health departments for people who died. Population-based tumor registries for assessing cancer deaths in Texas, Indiana, and Minnesota were not established until 1949, 1987, and 1988, respectively, and cannot be used in all areas for cancer-mortality studies covering much of the period of interest.

For some diseases, excess-mortality studies can be carried out with information on death certificates for persons in exposed and unexposed communities. However, death certificates are not reliable for detecting many of the diseases mentioned in the public meetings. Excess-mortality studies have other serious limitations, including lack of information on place of residence during the time of exposures, lack of information on the background characteristics of study populations that affect disease likelihood (National Center for Health Statistics 1994) and lack of information on other exposures that could cause the diseases being studied. For other nonfatal health outcomes, there are no standardized reporting mechanisms in place that would permit a scientific analysis. The long time since exposure occurred, the limitations of health information, the diversity of reported health concerns, and the variations in survival and residence of individuals argue against conducting an excess-mortality or excess-morbidity study.

Cancer was a frequently mentioned health concern at all public meetings. That could in part reflect the importance and prevalence of cancer in the U.S. population. Cancer (of all types) is the second leading cause of death and illness in the United States (American Cancer Society 1996). About one in 1 in 2 males and 1 in 3 females will develop cancer. Lung cancer will occur 1 in 12 males and 1 in 19 females. Smokers are 10-20 times more likely than nonsmokers to develop lung cancer. Breast cancer is a leading cause of cancer in women, affecting 1 in 8 women at some point in their lifetime. Some people who spoke at the public hearings felt that cancer was unusually common in their communities and wondered whether there had been a cluster of cancer deaths in their areas. Epidemiologists have investigated numerous reports of cancer clusters throughout the world. In most instances, if a higher than expected rate of cancer is observed, no cause can be established. The Centers for Disease Control reviewed 108 cancer clusters in 29 states and 5 other countries and noted that no clear cause could be found for any of the clusters (Caldwell 1990). In fact, cluster-based studies work best for outbreaks of infectious diseases, such as Legionnaire's disease or *Salmonella* infections, which have an acute onset in a well-defined population soon after exposure. The subcommittee has concluded that a cluster-based epidemiologic design would not be a sound approach to study the chronic-disease consequences of ZnCdS exposure.

The subcommittee also considered a prospective cohort study of children who attended the Clinton Street School in Minneapolis. Any study based on the Clinton Street School would have several methodological shortcomings. First, exposure status would need to be assigned on the basis of geographic residence, assuming that all children in a geographic area had the same exposure. Second, records that could be used to identify all the children who attended the school are not available. No records regarding the children's health status before and after the Army's tests are available. As previously noted, documenting changes over time is important to establish a temporal relationship between exposure and outcome. Furthermore, a small number of students are available for study; the subcommittee's power calculations highlight the large number of people needed for valid studies of selected health outcomes. Finally, information on potential confounders in this population (such as second-hand exposure to cigarette smoke or exposures related to indoor and outdoor air pollution) cannot be reliably assessed, because too much time has elapsed. Collectively, those serious limitations argue against further research involving this special population.

CONCLUSIONS

The subcommittee evaluated the feasibility of an epidemiologic study in relation to various methodologic issues. The evaluation is summarized in Table 7-1 with the subcommittee's rationale for determining the feasibility of future epidemiologic studies. As Table 7-1 reflects, only a crude estimate of exposure based on people's residence at the time of the Army's tests can be ascertained for affected communities. Although current health status could be determined for some, an epidemiologic study would not be able to determine scientifically whether ZnCdS exposure caused a disease, given the absence of reliable information on exposure, individual dose, or other potential confounders. Small increases in cadmium exposure or cancer risk attributable to cadmium from ZnCdS exposure would be indistinguishable from those related to background sources of cadmium and other risk factors for cancer. And a very large number of people would be needed for a study to determine statistically significant differences attributable to ZnCdS exposure.

In summary, the subcommittee has concluded that there are three major

TABLE 7-1 Methodologic Considerations in Assessing the Feasibility of an Epidemiologic Study of the Health Effects Associated with ZnCdS Exposure

Methodologic Consideration	Feasibility	Rationale
1. Is a well-defined population comprising clearly defined exposed and unexposed individuals available for study?	Yes	The Army monitored the dispersion of ZnCdS. Exposure could crudely be estimated on the basis of residence at the time of the tests; however, this ecologic approach is subject to misclassification bias with respect to exposure status.
2. Is the exposure capable of causing the suspected health outcomes?	No	Although lung cancer and kidney disease have been linked to occupational cadmium exposure, no known health effects have been associated with ZnCdS. Exposures (to cadmium from ZnCdS in a worse-case scenario) resulting from the Army's tests are believed to be too low to cause health effects. No comprehensive health records exist for people affected for the periods of interest. Even if health status were evaluated currently, individual health complaints could not be linked specifically to past ZnCdS exposure.
3. Is the other information needed for a well-designed epidemiologic study available?		
a. Can confounders be adequately addressed in the study?	No	No information is available on people with respect to known and potential confounders, such as cigarette-smoking, air pollution, diet, and other occupational exposures that might result in erroneous findings.
b. What sample size would be required to conduct a valid epidemiologic study?	No	At least 15 million exposed and 15 million unexposed individuals would be needed for valid assessment of such health concerns as lung cancer or kidney disease.

barriers to carrying out an epidemiologic study of the health effects of ZnCdS: lack of complete and accurate exposure data on individuals; inadequacies in data on health outcomes before, during, and after the periods of exposure; and, because of the low exposures, the requirement of a huge sample to detect any small increase in adverse health effects. Accurate measurement of individuals' doses is not possible, given the very small contribution of ZnCdS to the concentration of cadmium in the environment. Information on potential confounders also would be lacking.

The subcommittee concludes that an epidemiologic study of the affected populations is not feasible.

8

CONCLUSIONS AND RECOMMENDATIONS

THE SUBCOMMITTEE WAS charged to determine whether the exposure of the public to ZnCdS in many U.S. and Canadian locations, from the Army's dispersion tests, posed any health risks. To address this charge, the subcommittee reviewed the available toxicity data on ZnCdS and cadmium compounds, the input from the public concerning the health problems they reported, and exposure data to determine whether exposure to ZnCdS was related to any health effects in the exposed persons. The subcommittee also quantified the risk of potential health effects of ZnCdS exposures. The subcommittee's conclusions and recommendations are as follows.

INPUT FROM THE PUBLIC

People were outraged by being exposed to chemicals by the government without their knowledge or consent. The subcommittee did not address ethical and other social issues about the Army's dispersion tests; these questions are important, and the Army must develop a mechanism for addressing the public's sense of outrage, but these issues were beyond the subcommittee's charge and expertise.

TOXICITY AND RELATED DATA ON ZINC CADMIUM SULFIDE

Data on the toxicity of ZnCdS are sparse. Results of feeding studies conducted in dogs and rats suggest that ZnCdS is practically nontoxic following acute (single, high-dose) exposures. ZnCdS was not found to be a skin or eye irritant in rabbits.

The physical and chemical properties of ZnCdS are fairly well understood. It is soluble only in strong acids and probably not absorbed through the skin or gastrointestinal tract. Its lack of solubility suggests that it is highly unlikely that free cadmium ions would become bioavailable as a result of inhalation of ZnCdS; however, information is not available on whether ZnCdS might break down in the respiratory tract into more-soluble components, which could easily be absorbed into the blood. The information on physical and chemical properties and the limited toxicity data available suggests that ZnCdS is not likely to show systemic toxicity.

The subcommittee in its interim report (NRC 1995) recommended that the Army conduct studies to determine the bioavailability and inhalation toxicity of ZnCdS in experimental animals. This research will strengthen the database needed for assessing the risks of ZnCdS and lessen the need to rely on the use of a surrogate for toxicity information. The Army has begun the research recommended by the subcommittee, and the results are expected shortly. The subcommittee recommends that when the results of the research become available, they be reviewed by experts outside the Army to determine whether the subcommittee's conclusions are still valid or should be modified.

TOXICITY AND RELATED DATA ON SELECTED CADMIUM COMPOUNDS

Because the toxicity of ZnCdS is largely unknown, the subcommittee examined the toxicity and related data on the most-toxic component of ZnCdS, cadmium—CdS, an insoluble cadmium compound, for noncancer health effects and all cadmium compounds for cancer.

Bioavailability of cadmium is a key determinant of the toxicity of a cadmium compound. Cadmium compounds differ in their toxicity: compounds that are insoluble in vivo (and so not bioavailable), such as CdS, are less toxic than compounds that are soluble in vivo, such as cadmium chloride, $CdCl_2$. Because of similarities in physical and chemical properties and toxicity, the subcommittee believes that CdS should be considered a surrogate for ZnCdS for noncancer health effects. Results of several toxicity studies conducted in experimental animals indicate that CdS inhaled into the lung is not solubilized and absorbed into the blood for distribution to other organs. Therefore, the subcommittee concluded that sporadic exposure of the public to small amounts of ZnCdS should not result in any noncancer systemic toxicity (toxicity to the kidneys, bone, immune system, or reproductive and developmental system). The only organ considered to be potentially at risk because of exposure to particles of ZnCdS was the respiratory tract; the particles of ZnCdS were respirable and would be expected to deposit deep in the lungs. Results of several toxicity studies indicate that CdS has little toxicity in the respiratory tract. Rats exposed to CdS at 39,600 mg-min/m^3 had only a mild pulmonary response.

Inhaled cadmium has been shown in occupational studies and laboratory studies of animals to cause lung cancer but not cancer at other body sites. Cadmium exposures associated with increased lung-cancer risk in human and animal studies were to much higher concentrations for longer periods and involved more biologically soluble compounds than the exposures to cadmium from ZnCdS in the Army's testing program.

EXPOSURE ASSESSMENT

The subcommittee reviewed the Army's documents and estimated the potential inhalation doses of cadmium from ZnCdS that were achieved during its use in the Army's air dispersion tests. The four highest potential inhaled doses of cadmium were in Minneapolis (6.8 µg), Winnipeg (14.5 µg), St. Louis (24.4 µg), and Biltmore Beach, FL (390 µg).

Estimated average total daily intake from environmental and industrial exposures to cadmium from all media (soil, water, food, and air) of 12-84 µg in urban areas is greater than any daily exposures to cadmium resulting from the ZnCdS particles in the Army tests, with the exception of the test conducted at Biltmore Beach, an isolated beach area in Florida. It should be noted that the cadmium intake directly from air (that is, via inhalation) contributes very little to total cadmium intake in urban and industrial areas. At approximately half the Army's test sites, the highest monitored ZnCdS concentrations had concentrations of airborne cadmium (in the form of ZnCdS) that were above the estimated urban average daily airborne cadmium, but the subcommittee concluded that these short-term high concentrations would have minimal impact on total cadmium exposures which is mainly from water, food, and soil.

RISK ASSESSMENT OF ZnCdS EXPOSURES

The only organ system considered to be potentially at risk because of exposure to particles of ZnCdS was the respiratory tract; the particles of ZnCdS were respirable and would be expected to deposit deep in the lungs. Since little toxicity data is available on ZnCdS, the subcommittee based its assessment of the potential toxicity of ZnCdS for noncancer health effects on the toxicity of CdS. Results of several toxicity studies indicate that CdS has little toxicity in the respiratory tract. Rats exposed to CdS at 39,600 mg-min/m^3 had only a mild pulmonary response. Using that result and dividing by 10 to extrapolate from a lowest-observed-adverse-effect level to a no-observed-adverse-effect level by 10 to extrapolate from animal to human exposures, and by 10 to account for sensitive populations, one would not expect adverse noncancer health effects, even in sensitive populations, from exposure to cadmium in an insoluble form, such as CdS, at 39.6 mg-min/m^3 or 39,600 μg-min/m^3, or cadmium at 30,900 µg-min/m^3 (39,600 × 0.78 = 30,900 µg-min/m^3), or 513 µg of inhaled cadmium dose.

The four highest estimated inhaled cadmium doses were in Minneapolis (6.8 µg), Winnipeg (14.5 µg), St. Louis (24.4. µg), and Biltmore Beach, FL (390 µg). The potential doses of cadmium from ZnCdS tests did not exceed the level considered to be "safe" (513 µg) for noncancer effects at any test site.

The subcommittee also estimated maximal lifetime excess lung cancer risk for the locations with the highest cadmium exposures (i.e., Biltmore Beach, St. Louis, Winnipeg, and Minneapolis). Cancer risk estimates were based on the cadmium content of ZnCdS because no studies on the carcinogenicity of ZnCdS have been conducted and cadmium is the most-toxic component, so this approach is likely to overestimate the risk. For each test site, the location with the highest average reported air concentration of cadmium was used to calculate cancer risk. The risk estimates were multiplied by a factor of 10 to account for exposure to sensitive populations. The estimated upper-bound lung-cancer risks are small and ranged from less than 0.01×10^{-6} to 24.0×10^{-6}.

The vast majority of people were exposed to concentrations that were less than the highest by a factor of 2-10 or even more. Accordingly, their cancer risks would be lower by a factor of 2-10 or more. The subcommittee concluded that given the extremely low concentrations of ZnCdS (or cadmium from ZnCdS) in the air and the short duration of exposure, the lung-cancer risk, if any, is likely to be very low. Thus, it is unlikely that anyone in the test areas would have developed lung cancer from direct exposure to airborne releases of ZnCdS. On the basis of a review of the scientific literature, information presented at the public meetings, and an analysis of exposure information, the subcommittee concluded that the list of disorders reported by people living in areas where ZnCdS was released reflects diseases found in the general population and cannot be attributed to the ZnCdS releases.

FEASIBILITY OF EPIDEMIOLOGIC STUDY

The subcommittee evaluated the feasibility of an epidemiologic study in relation to various methodologic issues. Various study designs were considered in light of the information presented by members of the public attending the public hearings.

The subcommittee has concluded that there are three major barriers to carrying out an epidemiologic study of the health effects of ZnCdS: lack of complete and accurate exposure data on individuals; inadequacies in data on health outcomes before, during, and after the periods of exposure; and, because of the low exposures, the requirement of a huge sample to detect any small increase in adverse health effects. Accurate measurement of individuals' doses is probably not possible, given the very small contribution of ZnCdS to the concentration of cadmium in the environment. Information on potential confounders also would be lacking. The subcommittee concludes that an epidemiologic study of the affected populations is not feasible.

RECOMMENDATIONS

It is highly unlikely that free cadmium ions would become bioavailable to target organs as a direct result of inhalation of ZnCdS. However, information is not available on whether ZnCdS might break down in the respiratory tract into more-soluble components, which could be absorbed into the blood.

- The subcommittee recommends that the Army conduct studies to determine the bioavailability and inhalation toxicity of ZnCdS in experimental animals. This research will strengthen the database needed for risk assessment of ZnCdS and lessen the need to rely on the use of cadmium or cadmium compounds as surrogates for toxicity information.
- The subcommittee recommends that when the results of the research become available, they be reviewed by experts outside the Army to determine whether the subcommittee's conclusions are still valid or should be modified.

People were outraged at being exposed to chemicals by the government without their knowledge or consent.

- The subcommittee did not address ethical and other social issues about the Army's dispersion tests; these questions are important, and the Army must develop a mechanism for addressing the public's sense of outrage, but these issues were beyond the subcommittee's charge and expertise.

REFERENCES

AEHA (U.S. Army Environmental Hygiene Agency). 1994. Assessment of Health Risk: Corpus Christi, Texas, and Minneapolis, Minnesota. Health Risk Assessment Study 64-50-93QE-94. U.S. Army Environmental Hygiene Agency, Aberdeen Proving Ground, Edgewood, Md.

AEHA (U.S. Army Environmental Hygiene Agency). 1995a. Assessment of Health Risk: Fort Wayne, Indiana. Health Risk Assessment Study 39-26-0467-95. U.S. Army Environmental Hygiene Agency, Aberdeen Proving Ground, Edgewood, Md.

AEHA (U.S. Army Environmental Hygiene Agency). 1995b. Assessment of Health Risk: St. Louis, Missouri. Health Risk Assessment Study 39-26-0467-95. U.S. Army Environmental Hygiene Agency, Aberdeen Proving Ground, Edgewood, Md.

Ali, M.M., R.C. Murthy, and S.V. Chandra. 1986. Developmental and long-term neurobehavioral toxicity of low level in utero cadmium exposure in rats. Neurobehav. Toxicol. Teratol. 8:463-468.

American Cancer Society. Cancer Facts & Figures—1996. American Cancer Society, Atlanta, Ga.

ATSDR (Agency for Toxic Substances and Disease Registry). 1993. Toxicological Profile for Cadmium. Rep. TP-92/06. U.S. Department of Health and Human Services, Public Health Service, Agency for Toxic Substances and Disease Registry, Atlanta, Ga.

ATSDR (Agency for Toxic Substances and Disease Registry). 1994. Toxicological Profile for Zinc. Rep. TP-93/15. U.S. Department of Health and Hu-

man Services, Public Health Service, Agency for Toxic Substances and Disease Registry, Atlanta, Ga.
Baranski, B. 1985. Effect of exposure of pregnant rats to cadmium on prenatal and postnatal development of the young. J. Hyg. Epidemiol. Microbiol. Immunol. 29:253-262.
Baranski, B. 1987. Effect of cadmium on prenatal development and on tissue cadmium, copper, and zinc concentrations in rats. Environ. Res. 42(1):54-62.
Baranski, B., I. Stetkiewicz, K. Sitarek, and W. Szymczak. 1983. Effects of oral, subchronic cadmium administration on fertility, prenatal and postnatal progeny development in rats. Arch. Toxicol. 54(4):297-302.
Bhattacharyya, M.H., A.K. Wilson, E.K. Silbergeld, L. Watson, and E. Jeffrey. 1995. Metal-induced osteotoxicities. Pp. 465-498 in Metal Toxicology, R.A. Goyer, C.D. Klaassen, and M.P. Waalkes, eds. San Diego, Calif.: Academic.
Bodek, I., W.J. Lyman, W.F. Reehl, and D.H. Rosenblatt, eds. 1988. Environmental Inorganic Chemistry: Properties, Processes, and Estimation Methods. New York: Pergamon.
Bruce, W.R., and J.A. Heddle. 1979. The mutagenic activity of 61 agents as determined by the micronucleus, *Salmonella*, and sperm abnormality assays. Can. J. Genet. Cytol. 21:319-334.
Buckley, B.J., and D.J. Bassett. 1987. Pulmonary cadmium oxide toxicity in the rat. J. Toxicol. Environ. Health 21:233-250.
Bui, T.H., J. Lindsten, and G.F. Nordberg. 1975. Chromosome analysis of lymphocytes from cadmium workers and Itai-Itai patients. Environ. Res. 9:187-195.
Caldwell, G.G. 1990. Twenty-two years of cancer cluster investigations at the Centers for Disease Control. Am. J. Epidemiol. Suppl. 132:PS43-PS47.
Chen, J.J., R.L. Kodell, and D.W. Gaylor. 1988. Using the biological two-stage model to assess risk from short-term exposures. Risk Anal. 8:223-230.
Chisolm, J.C. and C.R. Handorf. 1985. Zinc, cadmium, metallothionein, and progesterone: Do they participate in the etiology of pregnancy-induced hypertension? Med. Hypotheses 17:231-242.
Chisolm, J.C., and C.R. Handorf. 1987. Increased absorption of and sensitivity to cadmium during late pregnancy: Is there a relationship between markedly decreased maternal cadmium binding protein (metallothionein) and pregnancy-induced hypertension? Med. Hypothesis 24:347-351.
Chung, J., N.O. Nartey, and M.G. Cherian. 1986. Metallothionein levels in liver and kidney of Canadians: A potential indicator of environmental exposure to cadmium. Arch. Environ. Health 41:319-323.
Cifone, M.G., A. Procopio, T. Napolitano, E. Alesse, G. Santoni, and A. San-

toni. 1990. Cadmium inhibits spontaneous (NK), antibody-mediated (ADCC) and IL-2 stimulated cytotoxic functions of natural killer cells. Immunopharmacology 20:73-80.

Davidson, C.I., W.D. Goold, T.P. Mathison, G.B. Wiersma, K.W. Brown, and M.T. Reilly. 1985. Airborne trace elements in Great Smoky Mountains, Olympic, and Glacier National Parks. Environ. Sci. Technol. 19(1):27-35.

Deaven, L.L., and E.W. Campbell. 1980. Factors affecting the induction of chromosomal aberrations by cadmium in Chinese hamster cells. Cytogenet. Cell. Genet. 26:251-260.

Deknudt, G., and A. Leonard. 1975. Cytogenetic investigations on leucocytes of workers from a cadmium plant. Environ. Physiol. Biochem. 5:319-327.

Denizeau, F., and M. Marion. 1989. Genotoxic effects of heavy metals in rat hepatocytes. Cell Biol. Toxicol. 5:15-25.

DHHS (U.S. Department of Health and Human Services). 1986. The Health consequences of Involuntary Smoking: A Report of the Surgeon General. Rockville, Md.: U.S. Department of Health and Human Services, Public Health Service, Centers for Disease Control, Center for Health Promotion and Education, Office of Smoking and Health.

Duggan, R.E., and P.E. Corneliussen. 1972. Dietary intake of pesticide chemicals in the United States (III), June 1968–April 1970. Pestic. Monit. J. 5:331-341.

Elinder, C.-G., T. Kjellström, B. Lind, L. Linnman, M. Piscator, and K. Sundstedt. 1983. Cadmium exposure from smoking cigarettes: Variations with time and country where purchased. Environ. Res. 32:220-227.

Elinder, C.-G. 1985. Cadmium: Uses, occurrence, and intake. In Cadmium and Health: A Toxicological and Epidemiological Appraisal. Vol. I. Exposure, Dose, and Metabolism. Effects and response. L. Friberg, C.-G. Elinder, T. Kjellström, G.F. Nordberg, eds. Boca Raton, Fla.: CRC Press.

Elinder, C.-G., T. Kjellström, B. Lind, and L. Linnman. 1976. Cadmium concentration in kidney cortex, liver, and pancreas among autopsied Swedes. Arch. Environ. Health 31:292-302.

Ellis, K.J., D. Vartsky, and I. Zanzi. 1979. Cadmium: In vivo measurement in smokers and nonsmokers. Science 205:323-325.

EPA (U.S. Environmental Protection Agency). 1981. Health assessment document for cadmium. EPA-600/8-81-023. Research Triangle Park, N.C.: U.S. Environmental Protection Agency, Environmental Criteria and Assessment Office.

EPA (U.S. Environmental Protection Agency). 1983. Updated Mutagenicity and Carcinogenicity Assessment of Cadmium. EPA-600/8-83-025F. Office of Research and Development, Office of Health and Environmental Assessment, U.S. Environmental Protection Agency, Washington, D.C.

EPA (U.S. Environmental Protection Agency). 1992. Respiratory Health, Effects of Passive Smoking: Lung Cancer and Other Disorders. EPA/600/6-90/006F. Washington, D.C.: U.S. Environmental Protection Agency, Office of Research and Development, Office of Air and Radiation.

Epstein, S.S., E. Arnold, J. Andrea, W. Bass, and Y. Bishop. 1972. Detection of chemical mutagens by the dominant lethal assay in the mouse. Toxicol. Appl. Pharmacol. 23:288-325.

Exon, J.H., and L.D. Koller. 1986. Immunotoxicity of cadmium. Pp. 339-350 in Cadmium. Handbook of Experimental Pharmacology, Vol. 80, E.C. Foulkes, ed. Berlin, Germany: Springer-Verlag.

Falck, F.Y., Jr., L.J. Fine, R.G. Smith, K.D. McClatchey, T. Annesley, B. England, and A.M. Schork. 1983. Occupational cadmium exposure and renal status. Am. J. Ind. Med. 4:541-549.

Favino, A., F. Candura, G. Chiappino, G., and A. Cavalleri. 1968. Study on the androgen function of men exposed to cadmium. Med. Lav. 59:105-110.

Fedorov, V.A., V.A. Gahshin, and Y.N. Korkishko. 1993. Solid-state phase diagram of the zinc sulfide-cadmium sulfide system. Materials Res. Bull. 28:59-66.

Friberg, L. 1950. Health hazards in the manufacture of alkaline accumulators with special reference to chronic cadmium poisoning. Acta Med. Scand. 138 (Suppl. 240):1-124.

Friberg, L., M. Piscator, G.F. Nordberg, and T. Kjellström. 1974. Cadmium in the Environment, 2nd Ed. Boca Raton, Fla.: CRC.

Friberg, L., Elinder, C.-G., T. Kjellström, and G.F. Nordberg. 1986. Cadmium and health, a toxicological and epidemiological appraisal. Vol. II. Effects and response. Cleveland, Ohio: CRC Press.

Funkhouser, S.W., O. Martinez-Maza, and D.L. Vredevoe. 1994. Cadmium inhibits IL-6 production and IL6 mRNA expression in a human monocytic cell line, THP-1. Environ. Res. 66:77-86.

Gartrell, M.J., J.C. Craun, D.S. Podrebarac, and E.L. Gunderson. 1986. Pesticides, selected elements, and other chemicals in adult total diet samples, October 1980–March 1982. J. Assoc. Off. Anal. Chem. 69:146-159.

Gasiorek, K., and M. Bauchinger. 1981. Chromosome changes in human lymphocytes after separate and combined treatment with divalent salts of lead, cadmium, and zinc. Environ. Mutagen 3:513-518.

Gaylor, D.W. 1988. Risk assessment: Short-term exposure at various ages. Pp. 173-176 in Phenotypic Variation in Populations: Relevance to Risk Assessment, A.D. Woodhead, M.A. Bender, and R.C. Leonard, eds. New York: Plenum.

Glaser, U., H. Kloeppel, and D. Hochrainer. 1986. Bioavailability indicators of inhaled cadmium compounds. Ecotoxicol. Environ. Safety 11:261-271.

Goyer, R.A. 1995. Toxic effects of metals. Pp. 691-736 in Casarett and Doull's Toxicology: The Basic Science of Poisons, 5th Ed, C.D. Klaassen, M.O. Amdur, and J. Doull, eds. New York: McGraw-Hill.

Grose, E.C., J.H. Richards, R.H. Jaskot, M.G. Menache, J.A. Graham, and W.C. Dauterman. 1987. A comparative study of the effects of inhaled cadmium chloride and cadmium oxide: Pulmonary response. J. Toxicol. Environ. Health 21:219-232.

Gunderson, E.L. 1995. Dietary intakes of pesticides, selected elements, and other chemicals: FDA total diet study, June 1984–April 1986. J. AOAC Int. 78(4):910.

Hatch, M., and D. Thomas. 1993. Measurement issues in environmental epidemiology. Environ. Health Perspect. 101(Suppl. 4):49-57.

Heinrich, U., L. Peters, H. Ernst, S. Rittinghausen, C. Dasenbrock, and H. König. 1989. Investigation on the carcinogenic effects of various cadmium compounds after inhalation exposure in hamsters and mice. Exp. Pathol. 37:253-258.

Henderson, R.F., A.H. Rebar, J.A. Pickrell, and G.J. Newton. 1979. Early damage indicators in the lung III. Biochemical and cytological response of the lung to inhaled metal salts. Toxicol. Appl. Pharmacol. 50:123-136.

Hill, A.B. 1965. The environment and disease: Association or causation. Proc. R. Soc. Med. 58:295-300.

Horiguchi, H., N. Makaida, S. Okamoto, H. Teranishi, M. Kasuya, and K. Matsushima. 1993. Cadmium induces interleukin-8 production in human peripheral blood mononuclear cells with the concomitant generation of superoxide radicals. Lymphokine Cytokine Res. 12:421-428.

Huel, G., R.B. Everson, and L. Menger. 1984. Increased hair cadmium in newborns of women occupationally exposed to heavy metals. Environ. Res. 35:115-121.

IARC (International Agency for Research on Cancer). 1993. Beryllium, Cadmium, Mercury, and Exposures in the Glass Manufacturing Industry. IARC Monographs on the Evaluation of Carcinogenic Risks to Humans, Vol. 58. Lyon, France: International Agency for Research on Cancer.

IPCS (International Programme on Chemical Safety). 1992. Environmental Health Criteria 134: Cadmium. Geneva: World Health Organization.

Jarup, L., C.-G. Elinder, and G. Spang. 1990. Lung Cancer Mortality in Cadmium Exposed Battery Workers [Abstract]. Presented at the Congress of Occupational Medicine, Montreal, September 1990.

Kanematsu, N., M. Hara, and T. Kada. 1980. *rec* assay and mutagenicity studies on metal compounds. Mutat. Res. 77:109-116.

Kastelan, M., M. Gerencer, A. Kastelan, and S. Gamulin. 1981. Inhibition of mitogen and specific antigen-induced human lymphocyte proliferation by cadmium. Exp. Cell Biol. 49:15-19.

Kazantzis, G., and B.G. Armstrong. 1982. A Mortality Study of Cadmium Workers in the United Kingdom. A Report to the International Lead Zinc Research Organization, New York.

Kazantzis, G., F.V. Flynn, J.S. Spowage, and D.G. Trott. 1963. Renal tubular malfunction and pulmonary emphysema in cadmium pigment workers. Q. J. Med. 32(126):165-192.

Kazantzis, G., T.H. Lam, and K.R. Sullivan. 1988. Mortality of cadmium-exposed workers, a five-year update. Scand. J. Work Environ. Health 14:220-223.

Kelman, B.J., B.K. Walter, G.E. Jarboe, and L.B. Sasser. 1978. Effect of dietary cadmium on calcium metabolism in the rat during late gestation. Proc. Soc. Exp. Biol. Med. 158(4):614-617.

Kjellstrom, T., P.E. Evrin, and B. Rahnster. 1977. Dose-response analysis of cadmium-induced tubular proteinuria. A study of urinary beta-2-microglobulin excretion among workers in a battery factory. Environ. Res. 13:303-317.

Kjellstrom, T., C.-G. Elinder, and L. Friberg. 1984. Conceptual problems in establishing the critical concentration of cadmium in human kidney cortex. Environ. Res. 33:284-295.

Kleinbaum, D.G. 1994. Logistic Regression: A Self-Learning Text. New York: Springer-Verlag.

Klimisch, H.-J. 1993. Lung deposition, lung clearance and renal accumulation of inhaled cadmium chloride and cadmium sulphide in rats. Toxicology 84:103-124.

Kodell, R.L., D.W. Gaylor, and J.J. Chen. 1987. Using average lifetime dose rate for intermittent exposures to carcinogens. Risk Anal. 7:339-345.

Koller, L.D. 1973. Immunosuppression produced by lead, cadmium, and mercury. Am. J. Vet. Res. 34:1457-1458.

König, H.P., U. Heinrich, and H. Kock. 1992. Effect of photocorrosion on cadmium sulfide suspensions applied in animal inhalation studies with CdS particles. Arch. Environ. Contam. Toxicol. 22:30-5.

Konz, J., and P. Walker. 1979. An assessment of cadmium in drinking water from a multi-media perspective. Report to U.S. Environmental Protection Agency by the Mitre Corporation, McLean, Va.

Kotsonis, F.N., and C.D. Klaassen. 1977. Toxicity and distribution of cadmium administered to rats at sublethal doses. Toxicol. Appl. Pharmacol. 41:667-680.

Laudanski, T., M. Sipowicz, P. Modzelewski, J. Bolinski, J. Szamatowicz, G. Razniewska, and M. Akerlund. 1991. Influence of high lead and cadmium soil content on human reproductive outcome. Int. J. Gynecol. Obstet. 36:309-315.

Lawson, R.N. 1966. A technique for identifying suitable veins for blood sampling in breast cancer. Can. Med. Assoc. J. 94:451-453.

Lawson, R.N., and L.L. Alt. 1965. Skin temperature recording with phosphors: A new technique. Can. Med. Assoc. J. 92:255-260.

Leighton, P.A. 1955. The Stanford Fluorescent-Particle Tracer Technique: An Operational Manual. U.S. Chemical Corps Research and Development Program. Contract DA-18-064-CML-2564. Department of Chemistry, Stanford University, Palo Alto, Calif.

Leighton, P.A., W.A. Perkins, S.W. Grinnell, and F.X. Webster. 1965. The fluorescent particle atmospheric tracer. J. Appl. Meteorol. 4:334-348.

Loeser, E. 1974. Cadmium Chloride, Cadmium Pigment Yellow, Cadmium Pigment Red: Subchronic Toxicological Investigations on Rats (3-Month Feeding Trial). Rep. 4639, Bayer, Friedrich-Ebert-Str. 217, D-5600 Wuppertal 1.

Loiacono, N.J., J.H. Grazino, J.K Kline, D. Popovac, X. Ahmed, E. Gashi, A. Mehmeti, and B. Rajovic. 1992. Placental cadmium and birthweight in women living near a lead smelter. Arch. Environ. Health 47:250-255.

Machemer, L., and D. Lorke. 1981. Embryotoxic effect of cadmium on rats upon oral administration. Toxicol. Appl. Pharmacol. 58:438-443.

MacMahon, B., and T.F. Pugh. 1970. Epidemiology: Principles and Methods. Boston: Little, Brown.

Mahaffey, K.R., P.E. Corneliussen, C.F. Jelinek, and J.A. Fiorino. 1975. Heavy metal exposure from foods. Environ. Health Prespect. 12:63-69.

Mailhes, J.B., R.J. Preston, Z.P. Yuan, and H.S. Payne. 1988. Analysis of mouse metaphase II oocytes as an assay for chemically induced aneuploidy. Mutat. Res. 198:145-152.

Mandel, R., and H.J.P. Ryser. 1981. The mutagenic effect of cadmium in bacteria and its synergism with alkylating agents [abstract]. Environ. Mutagen 3:333.

Mandel, R. And H.J. Ryser. 1984. Mutagenicity of cadmium in *Salmonella typhinurium* and its synergism with two nitrosamines. Mutat. Res. 138:9-16.

Mason, H.J. 1990. Occupational cadmium exposure and testicular endocrine function. Hum. Exp. Toxicol. 9:91-94.

Morgan, H., and J.C. Sherlock. 1984. Cadmium intake and cadmium in the human kidney. Food Addit. Contam. 1:45-51.

Mukherjee, A., A.K. Giri, A. Sharma, and G. Talukder. 1988a. Relative efficacy of short-term tests in detecting genotoxic effects of cadmium chloride in mice in vivo. Mutat. Res. 206:285-296.

Mukherjee, A., A. Sharma, and G. Talukder. 1988b. Effect of selenium on cadmium-induced chromosomal aberrations in bone marrow cells of mice. Toxicol. Lett. 41:23-29.

Murthy, G.K., U. Rhea, and J.T. Peeler. 1971. Levels of antimony, cadmium, chromium, cobalt, manganese, and zinc in institutional total diets. Environ. Sci. Technol. 5:436.

National Center for Health Statistics. 1992. Current Estimates from the National Health Interview Survey, 1992. Vital Health Stat. Ser. No. 10(189).

Nogawa, K., R. Honda, T. Kido, I. Tsuritani, Y. Yamada, M. Ishizaki, and Y. Yamaya. 1989. A dose-response analysis of cadmium in the general environment with special reference to total cadmium intake limit. Environ. Res. 48:7-16.

Nordberg, G.F., T. Kjellstrom and M. Nordberg. 1985. Kinetics and metabolism. Pp. 103-178 in Cadmium and Health: A Toxicological and Epidemiological Appraisal, Vol. 1, Exposure, Dose, and Metabolism, L. Friberg, C.-G. Elinder, and T. Kjellstrom, eds. Boca Raton, Fla.: CRC.

NRC (National Research Council). 1980. Recommended Dietary Allowances. 9th Rev Ed. Washington, D.C.: National Academy Press.

NRC (National Research Council). 1986. Environmental Tobacco Smoke: Measuring Exposures and Assessing Health Effects. Washington, D.C.: National Academy Press.

NRC (National Research Council). 1989. Biologic Markers in Reproductive Toxicology. Washington, D.C: National Academy Press.

NRC (National Research Council). 1991a. Environmental Epidemiology. Vol. 1: Public Health and Hazardous Wastes. Washington, D.C.: National Academy Press.

NRC (National Research Council). 1991b. Human Exposure Assessment for Airborne Pollutants: Advances and Opportunities. Washington, D.C: National Academy Press.

NRC (National Research Council). 1994. Science and Judgment in Risk Assessment. Washington, D.C.: National Academy Press.

NRC (National Research Council). 1995. Interim Report of the Subcommittee on Zinc-Cadmium Sulfide. National Research Council, Washington, D.C.

NRC (National Research Council). 1997. Toxicologic Assessment of the Army's Zinc Cadmium Sulfide Dispersion Tests: Answers to Commonly Asked Questions. Washington, D.C.: National Academy Press.

Nriagu, J.O. 1980. Cadmium in the Environment. Part I: Ecological Cycling. New York: John Wiley & Sons.

Nriagu, J.O. 1981. Cadmium in the Environment. Part II: Health Effects. New York: John Wiley & Sons.

Oberdörster, G. 1990. Lung clearance mechanisms of soluble and insoluble cadmium compounds. Pp. 221-224 in Environmental Hygiene, Vol. 2, N.H. Seemayer and W. Hadnagy, eds. Berlin, Germany: Springer-Verlag.

Oberdörster, G., and C. Cox. 1990. Kinetics of inhaled $CdCl_2$, CdO and CdS in rats and monkeys. In Cadmium 89. Proceedings of the Sixth International Cadmium Conference. Cadmium Association, London, and Cadmium Council, Greenwich, Conn.

Oberdörster, G., D.J. Guth, Y.H. Lee, and R.D. Mavis. 1985. Lung toxicity of cadmium: Importance of its chemical form. In Heavy Metals in the Environment, T.D. Lekkas, ed. Edinburgh, Scotland: CEP Consultants.

Oberly, T.J., C.E. Piper, and D.S. McDonald. 1982. Mutagenicity of metal salts in the L5178Y mouse lymphoma assay. J. Toxicol. Environ. Health 9:367-376.

Oldereid, N.B., Y. Thomassen, A. Attramadal, B. Oliasen, K. and Purvis. 1993. Concentrations of lead, cadmium, and zinc in the tissues of reproductive organs of men. J. Reprod. Fertil. 99:421-25.

Oldiges, H., D. Hochrainer, and U. Glaser. 1989. Long-term inhalation study with Wistar rats and four cadmium compounds. Toxicol. Environ. Chem. 19:217-222.

O'Riordan, M.L., E.G. Hughes, and H.J. Evans. 1978. Chromosome studies on blood lymphocytes of men occupationally exposed to cadmium. Mutat. Res. 58:305-311.

OSHA (Occupational Safety and Health Administration). 1989. Air Contaminants. Final Rule 29 CFR. Fed. Regist. 54(Jan. 19):2332-2983.

OSHA (U.S. Occupational Safety and Health Administration). 1992. Occupational Exposure to Cadmium. Final Rule. Fed. Regist. 57(Sept. 14):42102-42463.

Petering, H.G., H. Choudhury, and K.L. Stemmer. 1979. Some effects of oral ingestion of cadmium on zinc, copper, and iron metabolism. Environ. Health Perspect. 28:97-106.

Pond, W.G., and E.F. Walker, Jr. 1975. Effect of dietary Ca and Cd level of pregnant rats on reproduction and on dam and progeny tissue mineral concentrations. Proc. Soc. Exp. Biol. Med. 148(3):665-668.

Pott, F., U. Ziem, F.J. Reiffer, F. Huth, H. Ernst, and U. Mohr. 1987. Carcinogenicity studies on fibres, metal compounds and some other dusts in rats. Exp. Pathol. 32:129-152.

Prigge, E. 1978. Inhalative cadmium effects in pregnant and fetal rats. Toxicology 10:297-309.

Roels, H., R. Lauwerys, and A.N. Dardenne. 1983. The critical level of cadmium in human renal cortex: A reevaluation. Toxicol. Lett. 15:357-360.

Rohr, G., and M. Bauchinger. 1976. Chromosome analysis in cell cultures of

the Chinese hamster after application of cadmium sulfate. Mutat. Res. 40:125-130.
Rothman, K.J. 1993. Methodologic frontiers in environmental epidemiology. Environ. Health Perspect. 101(Suppl. 4):19-21.
Ruda, H.E., ed. 1992. Widegap II-VI Compounds for Opto-Electronic Applications. London: Chapman & Hall.
Rudzki, E., P. Rebandel, J. Stroinski, and K. Parapura. 1988. Reactions to cadmium. Contact Dermatitis 18:183-184.
Rusch, G.M., J.S. O'Grodnick, and W.E. Rinehart. 1986. Acute inhalation study in the rat of comparative uptake, distribution and excretion for different cadmium containing materials. Am. Ind. Hyg. Assoc. J. 47:754-763.
Saaranen, M., M. Kantola, S. Saarikoski, and T. Vanha-Perttula. 1989. Human seminal plasma cadmium: Comparison with fertility and smoking habits. Andrologia 21:140-145.
Saltzman, B.E., J. Cholak, and L.J. Schafer. 1985. Concentrations of six metals in the air of eight cities. Environ. Sci. Technol. 19:328-333.
Schroeder, H.A., and J.J. Balassa. 1961. Abnormal trace metals in man: Cadmium. J. Chron. Dis. 14:236-258.
Schroeder, H.A., and M. Mitchener. 1971. Toxic effects of trace elements on the reproduction of mice and rats. Arch. Environ. Health 23:102-106.
Schroeder, H.A., J.J. Balassa, and W.H. Vinton, Jr. 1964. Chromium, lead, cadmium, nickel, and titanium in mice: Effect on mortality, tumors, and tissue levels. J. Nutr. 83:239-250.
Schroeder, W.H., M. Dobson, and D.M. Kane. 1987. Toxic trace elements associated with airborne particulate matter: A review. JAPCA 37:1267-1285.
Shigematsu, I., S. Kitamaru, J. Takeuchi, M. Minowa, M. Nagai, T. Usui, and M. Fukushima. 1982. A retrospective mortality study on cadmium-exposed populations in Japan. Pp. 115-118 in Cadmium 81. Proceedings of the Third International Cadmium Conference. Cadmium Association, London, Cadmium Council, New York, and International Lead Zinc Research Organization, Research Triangle, N.C.
Shiraishi, Y., H. Kurahashi, and T.H. Yosida. 1972. Chromosomal aberrations in cultured human leukocytes induced by cadmium sulfide. Proc. Jpn. Acad. 48:133-137.
Smith, J.P., J.C. Smith, and A.J. McCall. 1960. Chronic poisoning from cadmium fume. J. Pathol. Bacteriol. 80:287-296.
Sorahan, T. 1987. Mortality from lung cancer among a cohort of nickel cadmium battery workers: 1946-84. Br. J. Ind. Med. 44:803-809.
Sorell, T.L., and J.H. Graziano. 1990. Effect of oral cadmium exposure during

pregnancy on maternal and fetal zinc metabolism in the rat. Toxicol. Appl. Pharmacol. 102:537-545.

Sutou, S., K. Yamamoto, H. Sendota, and M. Sugiyama. 1980. Toxicity, fertility, teratogenicity, and dominant lethal tests in rats administered cadmium subchronically. II. Fertility, teratogenicity, and dominant lethal tests. Ecotoxicol. Environ. Saf. 4:51-56.

Tang, X.M., X.Q. Chen, J.X. Zhang, and W.Q. Qin. 1990. Cytogenetic investigation in lymphocytes of people living in cadmium-polluted areas. Mutat. Res. 241:243-249.

Thun, M.J. 1990. Testimony to OSHA regarding proposed rule on cadmium. Presented at a public hearing to OSHA, June 7, 1990, Docket H 057A, Exhibit 33, the U.S. Occupational Safety and Health Administration, Washington, D.C.

Thun, M.J., T.M. Schnorr, A.B. Smith, W.E. Halperin, and R.A. Lemen. 1985. Mortality among a cohort of U.S. cadmium production workers—An update. J. Natl. Cancer Inst. 74:325-333.

Thun, M.J., A.M. Osorio, S. Schober, W.H. Hannon, B. Lewis, and W. Halperin. 1989. Nephropathy in cadmium workers: Assessment of risk from airborne occupational exposure to cadmium. Br. J. Ind. Med. 46:689-697.

Thun, M.J., C.-G. Elinder, and L. Friberg. 1991. Scientific basis for an occupational standard for cadmium. Am. J. Ind. Med. 20:629-642.

Tsvetkova, R.P. 1970. The effect of cadmium compounds on the generative function [in Russian]. Gig. Tr. Prof. Zabol. 14:31-33.

U.S. Army. 1959. Large-Scale Particulate Cloud Travel, Operation LAC. Abridged Report, Vol. 1. Tech. Rep. DPGTR 227. U.S. Army Chemical Corps Research and Development Command, Dugway Proving Ground, Dugway, Utah.

U.S. Army. 1977. U.S. Army Activity in the U.S. Biological Warfare Programs: 1962-1977, Vol. 2.

Wahlberg, J.E. 1977. Routine patch testing with cadmium chloride. Contact Dermatitis 3:293-296.

Watanabe, T., and A. Endo. 1982. Chromosome analysis of preimplantation embryos after cadmium treatment of oocytes at meiosis I. Environ. Mutagen. 4:563-567.

Watanabe, T., T. Shimada, and A. Endo. 1979. Mutagenic effects of cadmium on mammalian oocyte chromosomes. Mutat. Res. 67:349-356.

Weast, R.C., ed. 1985. CRC Handbook of Chemistry and Physics, 66th Ed. Boca Raton, Fla.: CRC.

Weast, R.C., M.J. Astle, and W.H. Beyer, eds. 1986. CRC Handbook of Chemistry and Physics, 67th Ed. Boca Raton, Fla.: CRC.

Webster, W.S. 1978. Cadmium-induced fetal growth retardation in the mouse. Arch. Environ. Health 33(1):36-42.

Whelton, B.D., M.H. Bhattacharyya, B.A. Carnes, E.S. Moretti, and D.P. Peterson. 1988. Female reproduction and pup survival and growth for mice fed a cadmium-containing purified diet through six consecutive rounds of gestation and lactation. J. Toxicol. Environ. Health 24(3):321-343.

White, M.E. 1977. Memorandum on Subject: Concerning Evaluation of Possible Adverse Effects of Zinc and Cadmium Sulfide Aerosol Exposure in White County, Arkansas Department of Health, Little Rock, Ark., June 8, 1977.

Wong, P.K. 1988. Mutagenicity of heavy metals. Bull. Environ. Contam. Toxicol. 40:597-603.

Zenick, H., L. Hastings, and M. Goldsmith. 1982. Chronic cadmium exposure: Relation to male reproductive toxicity and subsequent fetal outcome. J. Toxicol. Environ. Health 9:377-387.

Appendix A

Historical Background of the U.S. Biologic-Warfare Program

Historical Background of the U.S. Biologic-Warfare Program

BECAUSE OF CONCERN over possible use of biologic warfare (BW) by a foreign power against the United States and its allies, U.S. Secretary of War Henry Stimson in 1941 asked the National Research Council/National Academy of Sciences to investigate all phases of BW. In response, the Research Council appointed a committee of 9 prominent scientists, known as the War Bureau of Consultants (WBC) Committee, which conducted its work in the utmost secrecy.

The committee concluded in a classified report (NAS 1942) that BW was distinctly feasible and urged that appropriate steps be taken for defense against its use. (The report was declassified in 1988.) The report stated in part:

> The value of biological warfare will be a debatable question until it has been clearly proven or disproven by experience. The wide assumption is that any method which appears to offer advantages to a nation at war will be vigorously employed by that nation. There is but one logical course to pursue, namely, to study the possibilities of such warfare from every angle, make every preparation for reducing its effectiveness, and thereby reduce the likelihood of its use.

Secretary Stimson conveyed the committee's recommendations to President Roosevelt, who in May of 1942 authorized the secretary to create an organization within the Federal Security Agency to conduct the U.S. Biological Warfare Program so as to avoid public concern over America's vulnerability. The exchange of information on this subject with the United Kingdom and Canada, which had been inaugurated some months before, was continued, and provision was made for the interchange of biologic-warfare personnel among the 3 countries (U.S. Army 1977). The research and development work from these programs extended knowledge of the military use of pathogenic microorganisms, as well as identifying ways to defend against them. It was recognized that an effective program involving BW agents, weapons systems, and production could not be achieved without large-scale developmental operations. In 1942, the U.S. Army Chemical Warfare Service assumed the responsibility for a large-scale research and development program. Fort Detrick in Frederick, MD, was selected to carry out the biologic-warfare task.

BIOLOGIC-WARFARE TESTING

The policy of the United States concerning BW between 1941 and 1973, when the entire stockpile of offensive BW agents was destroyed as a consequence of a directive issued by President Nixon in 1969, was to deter its use against the United States and its allies and to retaliate if deterrence failed. Fundamental to the development of a deterrent strategy was the need for a thorough study and analysis of our vulnerability to overt and covert attack. The policy also required examination of the full range of retaliatory options and development of a retaliatory capability that used pathogenic agents. In short, the policy required extensive research and development to determine precisely our vulnerability, the efficacy of our protective measures, and the tactical and strategic capability of various BW agents and delivery systems.

In the beginning and continuing throughout the BW program, there was a paucity of scientific and engineering knowledge and principles related to the vulnerability of the United States and its allies to BW attacks. Vulnerability testing was conducted to provide information on the agents likely to be used, means of disseminating them, sizes of areas that could be attacked, environmental effects on agents, dispersion of agents, the obstructive effects of buildings and other structures and terrain on agents, the detection and identification of agents, U.S. areas and forces most likely to be attacked, the extent of damage possible, and physical and mathematical models that could be used as substitutes for live, open-air testing.

The pathogenic organisms considered most suitable as weapons were those of anthrax, brucellosis, tularemia, Q fever, and psittacosis. The tactical use of BW agents required the development of tables of munitions requirements (the quantity of material required to achieve a particular military objective) for the strategic use of BW agents against target cities. The Army recognized two major challenges. In the case of tactical use of BW agents, one of the greatest difficulties was in selecting conditions that would lead to successful results when logistically feasible amounts of material were used on comparatively small areas. As the potency of the newly developed agents increased, the objective became the distribution of small amounts of them over comparatively large areas. Thus, to use phosgene (a highly toxic chemical-warfare agent) effectively, it was desirable to identify and use conditions that would enable the effective use of several hundred pounds of phosgene per 100 yd^2, whereas some of the BW agents could be used in quantities of a few pounds per square mile.

The second challenge arose from the impossibility of field testing BW agents in test areas that are typical of the probable targets, such as selected cities of the former Soviet Union. If pathogens or extremely toxic chemicals were to be tested, they had to be handled in remote, isolated areas, but such areas generally are atypical of habitable regions. The Army wanted to know the effects of buildings, terrain, meteorologic conditions, and so on, on the dispersion of BW agents. To obtain some estimate of the amount of BW material required to meet particular objectives in given cities or industrial areas, an indirect approach would be needed.

Three approaches to the problem of estimating munitions requirements for the strategic use of BW agents against target cities were available. The first, which was investigated by the British, used wind-tunnel studies on city models and a small number of field tests within a city itself (Aanensen 1951). Important results were obtained, but it had not been established that the necessary range of meteorologic conditions could be reproduced within a wind tunnel, particularly the unique mesometeorologic conditions induced by the city itself. The second and third approaches were based on full-scale field experiments. The second approach was to construct a test target typical of known targets in an area resembling as nearly as possible typical terrain conditions and having typical meteorologic conditions. Because cities themselves modify meteorologic conditions to some extent, the target had to be large enough to induce such modifications. The third approach was to simulate the BW agent, rather than the target, and to run tests with the simulant in suitable areas. This approach was chosen after the first 2 were rejected—the first because it was inadequate and the second because it was too expensive. That left researchers with two more choices: choosing the simulant and selecting test cities or other appropriate test locations.

SELECTION OF SIMULANT MATERIALS

Every effort expended in open-air testing was first directed toward the use of biologic and nonbiologic simulants to obtain the necessary data for evaluation (U.S. Army 1977). Biologic simulants are defined as living microorganisms that are not normally capable of causing infection, that represent the physical and biologic characteristics of potential microbiologic agents, and that are considered medically safe to operating personnel and surrounding communities. Nonbiologic simulants are nonliving inert (usually inorganic) materials. They are chosen to resemble the size of BW agents for penetration in the respiratory tract; they are not themselves BW agents.

The Army used both biologic and nonbiologic simulants in dispersion tests. The biologic simulants used included *Serratia marcescens, Bacillus globigii, Bacillus subtilis,* and *Aspergillus fumigatus*; all 3 are considered to be generally nonpathogenic or of low virulence to normal populations. The nonbiologic simulants studied included zinc cadmium sulfide, ZnCdS; sulfur dioxide, SO_2; and soap bubbles.

Setting a munitions requirement (the quantity of material required to achieve a particular military objective) for a BW agent required researchers to consider, among other things, the toxicity and dose of an agent. Determination of the dose involves 2 variables: toxicity depends on the size of the agent, which determines where in the respiratory system the organisms or particles would lodge (Stanford University 1952, pp. 16-17); and deposition of organisms or particles in the alveoli is much more effective than deposition in the upper respiratory tract.

ZnCdS, a fluorescent pigment, was chosen as a simulant not just because of its detectable glow, but also because of its particle size. Its particle-size range, 0.5-3 μm, approximates that considered most effective in penetrating into the lungs (Stanford University 1952, p. 55). Other properties that made it desirable as a simulant were its economic feasibility; its lack of toxicity to humans, animals, and plants; its stability in the atmosphere; and its dispersibility. ZnCdS was considered to be "a simulant having no viability" or an "inert material" (Stanford University 1952, p. 22).

SELECTION OF TEST CITIES AND OTHER TEST SITES

To accomplish the Army's goal of estimating munitions requirements for the strategic use of BW agents against cities, the researchers considered as test areas North American metropolitan areas that most closely matched the meteorologic, terrain, population, and physical characteristics of the Soviet cities of interest, such as Moscow and Leningrad (Stanford University 1952, pp. 32-37). For winter conditions, it became apparent that the upper Mississippi Valley and adjacent areas presented the best possibility of matching some climatic characteristics and bracketing others. In this general geographic region, the following cities were considered: Oklahoma City, Kansas City, Omaha, Cincinnati, St. Louis, Chicago, Minneapolis, and Winnipeg. Of those, St. Louis and Minneapolis appeared to bracket the range of climatic values of interest to the best advantage (Stanford University 1952). The broad topographic requirement to be met was that of flat to rolling country at elevations generally below 1,000 ft above sea level. St. Louis and Minneapolis qualified in those respects. They are on one or more rivers—another desired feature.

In summer, that general area has considerably more precipitation than the Soviet area of interest. However, areas in North America with substantially the same precipitation as that in the Soviet area of interest were disqualified from further consideration if their terrain was mountainous, they were too high above sea level, or their rainfall was of a peculiar coastal nature (Stanford University 1952). To qualify as a site for summer tests, a city had to have cloudy and clear days and rainy and dry days within specified ranges of temperature and windiness. St. Louis and Winnipeg met the desired summer temperature range, and in other respects they also qualified as summer-test cities (Stanford University 1952).

Population was an additional consideration used in the decision to test in Minneapolis, St. Louis, and Winnipeg. The Stanford University (1952) study noted that "of the 82 Russian cities whose populations are greater than 100,000, only two have populations exceeding 1,000,000" (Stanford University 1952, p. 34). Minneapolis, St. Louis, and Winnipeg fit within that range, with populations of 500,000, 800,000, and 250,000, respectively (Stanford University 1952, p. 34). Population density is also an important factor, but this information was not available on the Soviet cities.

The structural characteristics of cities of the former Soviet Union could not be completely matched in American cities, but the essential features were found in various degrees. St. Louis, in particular, has structure heights in general not exceeding 3 stories. Industrial installations occur both in built-up areas and in isolated areas.

The presence of universities in Minneapolis, St. Louis, and Winnipeg was yet another advantage in that the universities provided an "ample pool of qualified personnel" to assist with field testing and data reduction (Stanford University 1952, pp. 40, 73). In Minneapolis, researchers also believed that the University of Minnesota laboratories might prove useful if the need arose (Stanford University 1952, p. 73); a substantial number of part-time personnel from the University of Minnesota were employed (Stanford University 1952, p. 34).

The cooperation of local officials and the local staffs of the U.S. Weather Bureau was important (Stanford University 1952, p. 40). Cooperation of the police departments and air-pollution control officers was enlisted to avoid problems with local officials in connection with the operations. To avoid disclosing the exact nature and purpose of the operations, a cover story was devised: city officials were told that the work was to obtain data pertinent to smoke screening of cities to prevent aerial observation (Stanford University 1952, p. 76).

APPENDIX A

Other test locations were selected to simulate other Soviet cities (such as San Francisco, CA and Panama City, FL), forests (such as Chippewa National Forest, MN), flatlands (such as Fort Wayne, IN, and Corpus Christi, TX), deserts, and unpopulated areas (Dugway Proving Ground, UT). Appendix B provides the reasons for selecting other locations.

ARMY TESTS WITH BIOLOGIC AND NONBIOLOGIC SIMULANTS

In tests with the biologic and nonbiologic simulants, public safety was stated to be the foremost consideration (Stanford University 1952). Organisms and materials that were considered by the scientific community to be safe were selected (U.S. Army 1977). A total of 160 tests using various simulants were conducted at 66 locations (both military and civilian targets) in the United States (including Alaska and Hawaii) and Canada. The specific dates, locations, and substances used are in an Army report (U.S. Army 1977). In the conduct of BW testing, specialized sampling and analysis techniques were used to determine the various parameters of the test requirements and the downwind travel distances. These were supplemented by rather complete meteorologic data gathering systems to define meteorologic conditions. Meteorologic conditions constituted an absolute controlling factor in whether a test was permitted to start or continue.

SOME EXAMPLES OF ARMY TESTS WITH BIOLOGIC SIMULANTS

The three most commonly used biologic simulants were *Serratia marcescens, Bacillus globigii,* and *Aspergillus fumigatus.* The Army released *S. marcescens* in eight tests to determine the vulnerability to enemy attacks (U.S. Army 1977). In addition, the Army conducted field testing with *B. globigii* and *Aspergillus. S. marcescens* and *Bacillus subtilis* were also released in San Francisco and New York subway systems, respectively, to study their dispersion.

ZINC CADMIUM SULFIDE DISPERSION TESTS

Dispersion tests were carried out in many cities and rural locations with particles of the nonbiologic simulant ZnCdS. A total of 34 tests involving the dispersion of ZnCdS particles were conducted in various U.S. and Canadian locations. Some of those tests involved simultaneous releases of ZnCdS and *Serratia marcescens* or *Bacillus globigii*.

During the 1950s and 1960s, Stanford University and the Ralph Parsons Company (both contractors for the U.S. Army Chemical Corps) conducted atmospheric-dispersion tests with ZnCdS particles in Minneapolis, MN; Corpus Christi, TX; St. Louis, MO; Fort Wayne, IN; and 29 other urban and rural locations in the United States and Canada. The tests were purportedly used to develop and verify meteorologic models for estimating the dispersion of aerosols in various environments. However, the real purpose was to obtain information that would be useful for estimating the potential dispersion of BW agents and determining munitions requirements for the strategic use of BW agents against selected cities of the former Soviet Union. The tests were designed to provide information on the dispersion of BW agents over a short distance (for example, within a city, such as Minneapolis), over many miles (Fort Wayne, IN), and over several thousand square miles (Large Area Coverage test).

Operation LAC, which took its name from "Large Area Coverage," was the largest test ever undertaken by the Chemical Corps. The test area covered the United States from the Rockies to the Atlantic, and from Canada to the Gulf of Mexico. The tests proved the feasibility of covering large areas (thousands of square miles) of a country with BW agents. Many scientists and officers had believed this possible, but LAC provided the first proof.

REFERENCES

Aanensen, C.J.M. 1951. Wind Tunnel Experiments of Diffusion in a Built-up Area. Porton Technical Paper 257. The Travel of Gas in a Built-up Area. Porton/TU 1206/340/51.

NAS (National Academy of Sciences). 1942. Report of the WBC Committee. (Declassified in 1988)

Stanford University. 1952. Behavior of Aerosol Clouds within Cities. Joint Quarterly Report Number One. July-September. Stanford University and Ralph Parsons Company. 78 pp.

U.S. Army. 1977. U.S. Army Activity in the U.S. Biological Warfare Programs, 1962-1977, Vol. II.

Appendix B

Summary of Doses and Concentrations of Zinc Cadmium Sulfide Particles from the Army's Dispersion Tests

Prepared by
Edmund Crouch

Cambridge Environmental, Inc.
Cambridge, Massachusetts

Summary of Doses and Concentrations of Zinc Cadmium Sulfide Particles from the Army's Dispersion Tests

I. INTRODUCTION

This document has been prepared for the Subcommittee on Zinc Cadmium Sulfide to summarize some estimates of concentrations and potential exposures (time-integrated concentrations) of zinc cadmium sulfide (ZnCdS) that were achieved during the use of this compound in certain air-dispersion tracer tests. The basic source materials were a subset of the references included in an appendix entitled "Zinc Cadmium Sulfide Testing Documents," which listed all known Army-sponsored ZnCdS tests. All reference numbers herein refer to the list for this appendix. The description of the aim of this report is best described by the scope of work:

A. Scope of Work (Cambridge Environmental Inc.)

Dr. Bakshi has provided Cambridge Environmental Inc. with approximately 50 documents (see the reference list for this appendix), most of which contain details of various U.S. Army-sponsored tests using ZnCdS tracer (some provide details of multiple tests in one or more series).

Cambridge Environmental, Inc., reviewed these documents and abstracted summaries of them, providing the following details for each test or series of tests whenever such details were available in the documentation:

>Name of test;
>Reference;
>Principal object of test;
>Number of and naming conventions for releases;
>A summary of the test conditions (method of release, location of release, mode of release, release vehicle, sampling station information, and so forth);
>Test material;
>Place of release;
>Communities that were affected by measurable concentrations of tracer;
>Date of releases;
>Time of day of releases;
>Period of time during which releases occurred;
>Distance of releases to affected communities;
>A summary of weather conditions during the releases;
>Total quantities of ZnCdS released;
>Total quantities of any other materials co-released;
>Maximum time-integrated concentration (microgram-minutes per cubic meter) due to each release;
>Maximum time-integrated concentration due to each release in any populated area;
>Maximum concentrations during the releases;
>Maximum concentrations during the releases in any populated area;
>Other comments.

Not all such details were available in all cases, although certain details always appear to have been provided. In all cases examined, measurements and estimates were made of time-integrated concentrations at various locations.

APPENDIX B

The two measurements of exposure—time-integrated concentration and concentration—are those required for estimating risks using the standard paradigms. In all cases, the release times were relatively short (<24 h), so the maximum concentration was compared with a short-term safety standard. Estimates of lifetime cancer risk are generally made using

$$R = U \frac{1}{T} \int C \, dt$$

where R is the lifetime risk estimate, U is the unit risk (m³/µg), T is a standard lifetime (70 years), C is the time-varying concentration, and t is time. This estimate involves the time-integrated concentration directly.

B. MISCELLANEOUS NOMENCLATURE

Throughout this document, various abbreviations and standard nomenclatures are used. The most important of these are the terms used to describe the two measurements of exposure or potential exposure evaluated. These two terms are

Concentration The mass of ZnCdS tracer material per unit volume of air, averaged over a relatively short period (generally 10 s to 2 h, depending on the experiment). All concentrations are reported here in units of microgram per cubic meter.

Exposure The time integral of the instantaneous concentration of ZnCdS at a single point of measurement, reported in microgram-minutes per cubic meter.

All the documents examined use the nomenclature "dose" to represent the time-integral of the concentration measured in particles of ZnCdS per unit volume (usually measured as particle-minutes per liter).

An attempt has been made to distinguish between concentrations and exposures measured in (humanly) populated areas and in unpopulated areas, although this attempt necessarily is limited by the available information.

Individuals, particularly those involved in the experiments, might have been subjected to the concentrations or exposures listed as being in unpopulated areas; in some cases, no one might have been subjected to the concentrations and exposures listed as being in populated areas.

The material of interest is fluorescent ZnCdS. This nomenclature is used for convenience only. The material used should probably be considered an alloy of zinc and cadmium sulfides, with the fluorescence color determined by minor additions of other elements. The stoichiometry is approximately $Zn_{0.8}Cd_{0.2}S$, but all masses used in this document refer to the total mass of the ZnCdS—no correction to obtain a mass of cadmium (the principal component of concern to the subcommittee) has been applied.

Some of the abbreviations used in the following summaries are

ppg	particles per gram	BG	*Bacillus globigii*
MMD	mass mean diameter	SM	*Serratia marcescens*
QR	quarterly report	C2	*Aspergillus fumigatus*
BMR	bimonthly report	Conc.	concentration
JQR	joint quarterly report	Max.	maximum
SAL	Stanford Aerosol Laboratory	Exp.	exposure

2. SUMMARY OF RESULTS

The entries in Table B-1 are the following:

Ref.	Reference number in the listing of reports
Place	Location of the releases
Name	Any known name given to the operation in which releases occurred
Start date	Date of first release
End date	Date of last release

TABLE B-1 Exposure Data on ZnCdS Dispersion Tests

						Populated Areas			Unpopulated Areas	
Ref.	Place	Name	Start Date	End Date	No. of Releases	Total Quantity of ZnCdS, kg	Max. Exp., μg-min/m³	Max. Conc., μg/m³	Max. Exp., μg-min/m³	Max. Conc., μg/m³
							Approximate area affected, square miles			
2	Camp Cooke, Calif.		1955		39	2.3	<173 <10	<7 <10	60,136 <1	55,650 <1
			Releases not at Dugway Proving Ground							
8	N. Carolina, S. Carolina, Georgia,	DEW I	03/26/52	04/21/52	5	630	98 <100,000	0.34 <10,000		
13	Corpus Christie, Tex.	WINDSOC	08/13/59	02/22/60	13	~1,600	? <100,000	? <10,000		
16	Oklahoma		06/04/62	06/16/62	9	204	39 <1,000	0.75 <1,000		
16	Texas		06/24/62	06/29/62	9	204	36 <1,000	0.82 <1,000		
16	Washington		10/02/62	10/21/62	9	204	6.7 <1,000	0.3 <1,000		
16	Nevada		10/31/62	11/05/62	8	181	23 <1,000	1.6 <1,000		
17 & 43	St. Louis		05/27/63	03/17/65	42	984	7,400 <10	40 <10		
19	Chippewa National Forest, Minn.		01/25/64	08/07/64	24	330	1,620 <1,000	3 <1,000	2,779 <100	9.1 <100
20	San Francisco		03/25/64	04/23/67	18	27.75	<1,900 <10	<170 <1	1,900 <10	170 <1
22	Fort Wayne, Ind.		02/02/64	02/04/66	75	~1,650	410 <1,000	<20 <1,000		
24	Oceanside, Calif.	Onshore	06/23/67	07/17/67	45	237	694 <1	1,741 <1		
		Offshore releases					149 <1,000	1.3 <1,000		
27	Pack Forest, Wash.		10/15/68	09/05/69	33	1.7			? <0.1	? <0.1
28	Green Brier Swamp, Md.	MATE	08/01/69	10/29/69	111	2.7	<42 <1	~0.03 <1		
30	Camp Detrick, Md.	SELTZER	02/18/53	02/24/53	4	0.022	71 <10	9 <1	785 <1	93 <1

TABLE B-1 (Continued)

						Total	Populated Areas		Unpopulated Areas	
							Max. Exp., μg-min/m^3	Max. Conc., μg/m^3	Max. Exp., μg-min/m^3	Max. Conc., μg/m^3
Ref.	Place	Name	Start Date	End Date	No. of Releases	Quantity, kg	Approximate area affected, square miles			
30	Biltmore Beach	WHITEHORSE	03/24/53	05/02/53	12	9.7	150,000 <0.1	4,800 <0.1		
33	Dallas		04/01/61	08/31/61	37		? <1,000	? <1,000		
35	St. Louis		01/19/53	10/18/53	35		2,000 <10	340 <0.1		
35	Minneapolis		05/20/53	06/23/53	102	7.9	2,600 <10	300 <0.1	17,200	1,100
35	Winnipeg		07/09/53	08/01/53	36	5.8	5,600 <10	1,000 <0.1	920 <1	130 <1
36	Stanford University, Calif.		10/15/47	10/15/47	1	0.00083	5	2		
37	Palo Alto, Calif.		03/10/50	03/14/50	2	0.976	2.4 <100	0.5 <100		
37	San Francisco		10/20/50	10/27/50	6	22.44	436 <100	15 <100		
41	Palo Alto, Calif.		01/26/62	11/16/62	28	1.4	1,676 <0.1	? <0.1		
Releases at Dugway Proving Ground										
3	Dugway		05/04/53	06/03/53	2	5.534			<77	
4	Dugway		01/21/54	03/14/54	4	0.0348			<2,000	
6	Dugway		05/18/55	05/18/55	2	0.0424			<6,00	
18	Dugway		05/17/63	08/15/63	9	29.6			1,044	
29	Dugway	GOOF	08/23/55	11/01/55	5	0.8	<409		14,300	
31	Dugway		04/03/58	04/22/58	4	0.0534				
32	Dugway		02/70 to 03/70		6	0.21			< 137,000	
37	Dugway		07/01/50	08/04/50	9	8.244	0.03		<80	0.5

Total
quantity — Quantity released in all known releases, including corrections for missing data

Number of
releases — Total number of known releases of ZnCdS

Populated
areas — Concentrations or exposures in populated communities, or, for Dugway, at the limits of the measurement array

Unpopulated
areas — Concentrations or exposures in unpopulated areas

Max. Exp. — The maximum exposure (time integral of concentration) of ZnCdS measured

Max. Conc. — The maximum concentration of ZnCdS measured or estimated

Approx. area
affected — An order of magnitude estimate of the area affected within an order of magnitude of the maximum concentration or exposure.

See the summary reports that follow for important caveats on all the reported values. In particular, all exposures are likely to be accurate to a factor of 2 at best, all concentrations will be substantially less accurate than that, and the magnitude of the values given often depends strongly on the placement of the sampling devices. All exposures and concentrations represent the total mass of ZnCdS present. (There is no correction to the mass of only cadmium.)

Exposure and concentration values are not directly comparable between experiments because of the different areas affected and the different placement of sampling points. To give an idea of the variation, the table includes (for the non-Dugway releases) some order of magnitude estimates for the areas affected within approximately an order of magnitude of the maximum concentrations or exposures reported. The areas given are approximate upper bounds on the areas affected, rounded up to the nearest power of 10.

A missing entry generally indicates a lack of, or irrelevance of, a measurement or estimate for that entry (e.g., exposures to unpopulated areas are irrelevant for releases made over populated areas).

In Chapter 5, Table 5-1 is derived from Table B-1. To obtain the cadmium dose, the maximal exposure of ZnCdS (expressed as $\mu g\text{-min}/m^3$ as shown in Table 5-1) is multiplied by 0.0166 m^3/min (the volume of air inhaled by an active person in one minute). The product is then multiplied by 0.156 (the mass fraction of cadmium in ZnCdS). The corresponding cadmium doses are 6.8 µg in Minnesota, 14.5 µg in Winnipeg, 24.4 µg in St. Louis, and 390 µg in Biltmore Beach.

APPROACH AND METHODS

A. Basic Approach

An initial scan of the approximately 50 documents available showed that there were too many and too varied a set of experiments to be able to perform complex or even simple analyses beyond those performed by the original experimenters and reported in the references. As far as possible, this report therefore contains simple abstracts of the most relevant data from the references, with as little interpolation as possible, and with only the simplest extrapolations.

The original researchers usually provided substantial documentation. A much more thorough job of reconstruction of exposures would in most cases be possible with substantially more resources, but, in view of the orders of magnitude of the exposures, such a reconstruction would probably not add much value for the committee.

B. Purpose of Original Experiments

In the references examined, there were two principal purposes for the experiments performed:

- Experimental confirmation of the practicality of dispersion of biologic warfare agents, empirical observations of the ranges and exposures that might be achieved, and obtaining correlations between the dispersal of biologic agents and inert aerosols. The fluorescent tracer was used as safe indicator of the potential extent of spread of biologic organisms under similar meteorologic conditions.

- Tracer tests in the atmosphere, to determine how the atmosphere behaves and derive correlations relating the dispersion of gases and aerosols to observable meteorologic conditions.

Of course, if correlations could be obtained between dispersal of biologic organisms and inert aerosols, the second purpose would also serve the first, although the second also has much wider application.

c. THE TYPICAL EXPERIMENT

The fluorescent-particle (FP) tracer technique is extensively described in Ref. 48. The idea is to release into the atmosphere a very large number of small particles as an aerosol that then disperse. The concentration of these particles after dispersion downwind is then measured by counting the individual particles collected by a collection device that samples a known volume of air. ZnCdS, originally developed as a fluorescent paint pigment, was found to be suitable for this application. The useful particle size range is about 0.5 to 5 μm, with the lower end limited by the detectability of the particles, and the upper end limited by the rate of fall-out from the air. The typical experiment used particles with a mass mean diameter of 1 to 2 μm, and an effective dispersal rate of around 1×10^{10} particles per gram (ppg).

It was found that the best available fluorescent ZnCdS particles could be reliably detected by eye under an optical microscope (with ultraviolet illumination to excite the fluorescence) at sizes down to around 0.3 μm diameter, allowing measurement of concentrations by counting individual particles collected on filters or other collection devices. Background counts in most areas were very low. No automatic particle-counting methods were described in the references examined. All particle counts were apparently obtained by human microscopic observations, although mechanical methods to improve the efficiency of sample preparation and presentation to the human observer are mentioned.

In a typical experiment, grams to kilograms of ZnCdS powder would be released over periods ranging from seconds to hours using an air-driven dispenser designed to efficiently produce an aerosol cloud. The dispenser would be mounted at a particular location (point source), or on the back of a moving vehicle (line source)—trucks, aircraft, and boats were used. The method of dispersion affected the dispersion efficiency, a measure of how much clumping of the particles occurred during release. Typical particles-per-gram values for particular manufacturers' lots of ZnCdS were apparently[1] obtained by a standardized test involving the dispersal of a small quantity of the material in an enclosed chamber. The particle count would then be obtained by sampling the aerosol with a filter-type sampler. The dispersion efficiency in an experiment is then the ratio of the particles per gram achieved in the experiment to the standardized value. A blending agent was typically used to improve the flow characteristics of the ZnCdS powder and reduce clumping, but environmental conditions could obviously affect the dispersion efficiency attained.

Downwind of the release point or line, sampling stations were set up at locations of interest. These sampling stations sampled the air by drawing a known volume flow rate (typically 6 liters (L)/min) through a filter (an approach that collects practically 100% of the particles on the filter) or by using a Rotorod collector. The latter consists of an H shaped wire (0.016 × 1/16 in.) with vertical members approximately 2.5 in. long, and crossbar 4.75 in. long, rotated upright about the center of the cross-bar. Particle collection is by impact on the 0.016-in.-wide leading edge of the vertical members of the H, which were coated with silicone (vacuum) grease to improve collection efficiency. The effective volume swept by the

---

[1]This description is inferential because no detailed description of the test method was located in the available material.

Rotorod was about 40 L/min, and collection efficiencies (which were calibrated against filter collectors) ranged from 30% to 80%, depending on the ZnCdS particle size distribution. Two independent samples could be obtained on each Rotorod by reversing the direction of rotation and using the other edge of the wire to obtain the second.

Sampling was typically started just before the release, and continued for a period lasting longer than the expected time during which any cloud of particles would be expected to be present. In some cases the time course of the cloud of aerosol was followed by using a sequential drum sampler, a type of filter sampler with a set of filters that could be automatically sequenced mechanically.

Filters and Rotorods would then be taken to a laboratory, and the collected particles counted. Reported results were always given in terms of exposure—typically as particle-minutes per liter.

The advantages of the FP tracer technique include its very high sensitivity, and its applicability over a wide range of scales. Statistically reliable particle counts require only 40 detected particles (for 15-20% accuracy), and with a collection volume flow rate of 6 L/min, this corresponds to an exposure of around 6 particle-min/L. At a count of 1×10^{10} ppg, this corresponds to 0.6 µg-min/m^3, or an average of 0.04 µg/m^3 for a 15-min sampling period. The technique was applied over scales that ranged from almost backyard to multistate.

D. Methods Used in This Document to Estimate Exposures and Concentrations

1. Exposures

All the references examined reported exposures in particle-minutes per liter or gave tabular data of raw particle counts (e.g., particles per filter). These values were, in general, taken at their face values (in one case (Ref. 19) a clearly incorrect description of tabular data was corrected). Reported values of particle-minutes per liter had generally been corrected for the collection efficiency of the sampling device, and these corrections were not questioned. Where only raw particle counts were given, the collection efficiency was also available, so that a correction could be applied for collection efficiency. The Dugway Proving Ground tests often gave only raw count data, but in the references examined these always used filter-type collectors with effectively 100% collection efficiency. In some of these cases, the sampling rate for the filter sampler was not given, and was assumed similar to the sampling rate in previous or subsequent experiments. Details are given in the individual summaries that follow.

The conversion from particle-minutes per liter to microgram-minutes per cubic meter (the unit of exposure used throughout this document) requires a particles-per-gram measurement for the dispersed aerosol in each experiment. Where this value was provided in the references, it was used, taking account also of any cited or cross-referenced dispersion efficiency for the device used in the particular experiment.[2] Where no dispersion efficiency was cited, but only a particles-per-gram count was given, that particles-per-gram count was generally taken to be the effective particles-per-gram count for the experiment (i.e., incorporating the dispersion efficiency), provided that this interpretation was consistent with the language of the reference and the result was within a reasonable range. Where no particles-per-gram count or dispersion efficiency was given, generic estimates based on the ZnCdS lot number or supplier and other experimental results were used to estimate the effective particles per gram of the dispersed aerosols. More details are given where necessary in the individual summaries that follow.

The estimates of particles per gram for the dispersed aerosols is necessarily approximate, and applying this estimate to the aerosols that were collected is doubly so. For these reasons, the exposure estimates are not considered accurate to better than a factor of two even in the best circum-

---

[2]The dispersion efficiency depends on the device, the vehicle from which the dispersion is taking place, and other conditions of the dispersion. Some full-scale tests of dispersion efficiency were done for some of the aerial line-source releases, using the vertical grid at Dugway Proving Ground.

APPENDIX B   135

stances. All exposure estimates in this document have thus been scaled (to microgram-minutes per cubic meter) to take account of both dispersion and collection efficiencies.

Exposures were often measured at many points, and the exposures varied substantially with location—particularly with distance from the sources. All exposure estimates given in this document correspond as closely as possible to the maximum exposures measured at any given point in any particular experiment. Where there were multiple releases, the exposures due to each release were summed at each measurement point, and the value abstracted here is the maximum of those sums. In a few cases, documented in the summaries that follow, it was impossible[3] to obtain the sum of the exposures at every measured point. In such cases, the best estimate available, or a conservative overestimate, for the maximum sum at any given point was used instead (e.g., in some cases, the maximum exposures—at different points—from each release were summed to obtain an overestimate of the maximum that could occur at any point).

In several cases, no tabular data were available, and it was necessary to estimate maximum exposures from contour data plotted on maps. Although individual measurement points were always also plotted, such individual points were often unreadable with the copies of the maps available. Because the contours plotted were widely spaced (often only in decades of particle-min per liter) and not completely consistent from map to map (some maps interpolated extra contours between some decades of particle-min per liter) and because the legends were often unreadable, this technique necessarily introduced further errors. The combined errors of reading from maps could amount to a factor of 5 in the exposure estimate in some cases.

The convention of using the highest measured value results in substantial

---

[3]Either the data were not available in the reference or obtaining the sum was logistically impractical with current resources because of the large quantity of data available.

variability due to the selection of sampling points. Exposures might vary dramatically with distance from a source, particularly from ground-level point or line sources, and particularly at small distances from the source. However, this approach was adopted because it requires the minimum extrapolation from the reported data. As mentioned in the Introduction, it would probably be possible in many cases to perform modeling to obtain a more standardized value for exposure, but such an approach would require resources beyond those available for this document.

Where appropriate and possible, separate estimates have been made for exposures in populated and unpopulated areas. Unpopulated areas were defined as areas not containing any buildings or street layouts indicating a built-up area on contemporary maps (if they were provided in the original reference) or on a modern computerized street map. "Unpopulated" areas included Dugway Proving Ground, the middle of a national forest, and the middle of a desert area. Caveats about such interpretations are provided in the individual summaries that follow.

II. CONCENTRATIONS

The methods used in these references did not allow direct measurements of concentrations, so that some inferences are required to provide estimates of concentration. The approach generally was to use as little extrapolation as possible, so that the concentration estimates are based as closely as possible on the measurements. However, the uncertainties in maximum concentrations are necessarily substantially larger than those in exposures.

For the purposes of this committee, the instantaneous concentration is of little consequence, since all concentrations are sufficiently low that no acute effects are expected under any circumstances (see Ref. 48 in Section 4.2, for example, where a short discussion of potential toxicity is abstracted from Ref. 48). Thus, whenever time-resolved data from an experiment were presented in the references, those data were used to estimate average concentrations over the resolution period of the data, even though the resolution varied from 10 s to 6 min.

Appendix B

The varied data collection and presentation methods given in the references necessitated varied approaches to estimating concentrations. The following techniques, or combinations of them, were used in different cases to estimate the maximum concentration that occurred during any single release at any measurement point. The methods used in each case are mentioned in slightly greater detail in the summaries that follow.

• If time-resolved data were available, those data were used. Time resolution varied from 10-s to 6-min averages, and those data were obtained with drum samplers. Where the time-resolved data did not coincide with the maximum exposure, the time-resolved data were adjusted upwards in proportion to relative exposures. This technique assumes a similar concentration-versus-time variation at both such points.

• For extended period point sources, the release periods were used to estimate the time the aerosol cloud could be present at any particular sampling point, and the exposures measured at those sampling points divided by the time periods to obtain an average concentration.

• For some instantaneous point sources, measured plume sizes allowed an estimate of the cross-wind standard deviation (σ_y) for an assumed gaussian shape. A similar standard deviation ($\sigma_y = \sigma_x$) was assumed for the downwind direction. Combining an assumption of downwind gaussian shape with the measured wind speed then allows an estimate for the peak concentration.

• For short releases from point or line sources, a combination of the two previous methods was applied. The release period was increased by twice the estimated downwind standard deviation in time (standard deviation in space divided by wind speed) to estimate the period the aerosol cloud might be present, and the average concentration obtained by dividing exposure by this period.

• Where no time-resolved data or aerosol-cloud-size data could be estimated from the measurements, but meteorologic data were available, the cloud was assumed to be gaussian in time, and the standard deviation in time of the concentration estimated by dividing σ_x by the wind speed, obtaining σ_x by assuming $\sigma_y = \sigma_x$, and estimating the former σ_y using the meteorologic conditions (using the standard estimates used in all EPA air-dispersion models).

- If no other methods could be used and some measurements of plume spread and wind speed were available, a standard line-source dispersion model was used for long-distance plume travel under inversion conditions from a line source. The measured value of σ_y was matched to the model to select the stability class, or the total exposure was matched to the measured exposure, and the model then provided an estimate of maximum concentration.

These approaches do not exhaust all possibilities used, because almost all references required a slightly different approach. For example, some methods used can be summarized as follows:

Reference	*Method*
No. 2	Cloud widths, wind speed, combination approach.
No. 5	2-h averages (samples).
No. 16	Standard dispersion model from line source, assume gaussian time shape and $\sigma_x = 1,000$ m (C stability at relevant distance).
No. 17	Reference shows concentrations—choose largest.
No. 18	15-min averages given.
No. 22	Approximate matching of measured σ_y and total exposure, using dispersion programs.
No. 24	Gaussian plume shape, total exposure, measured σ_y for point sources; total exposure, measured plume speed and duration for line sources.
No. 28	Rough empirical correlation provided in reference, plus total mass released.
No. 30	Total exposure, plume release time plus estimated σ_x.

Appendix B

III. Areas Affected

An estimate of the area affected was not included within the original scope-of-work, and much less effort has been devoted to these estimates, which are provided only in the summary table in Section 2 for the non-Dugway releases. The estimates of areas affected have been obtained by roughly estimating the length by width of the measured or inferred plumes in each experiment (or the angle and radius affected for point sources). The area so obtained (in square miles) was rounded up to the nearest power of 10, and entered in the table with a "less than" (<) indication. Where plumes went in substantially different directions in different releases, the area affected listed in the "Exposure" column in the table might be larger than that in the "Concentration" column, the latter indicating the size of individual plumes. The "size" of an aerosol plume is fairly arbitrary, since it depends on the definition given to the edge of the plume—much of the material in these plumes could travel over thousand mile distances if not rained out. As a rough approximation, wherever an explicit bound was needed, the distance to a concentration or exposure approximately 10 times lower than the maximum was used.

There are some cases in which a rough estimate of the plume size can be made (by using details of the experimental design), even though no estimate can be made of the concentration or exposure.

3. Miscellaneous Notes

A. References That Are Not Evaluable or Where More Data Are Needed

These references are probably available.

No. 1   Central Alaska. Volume 2B available. Need Volumes 1 and 2A for any useful report. No estimate made.
No. 13 Only 18 pages of more than 73 are available. The remainder, particularly the appendix, is needed. No estimates made.

No. 17 A short paper presentation to the annual meeting of the Air Pollution Control Association. Should be more information available somewhere. Only graphs with some particle concentrations and particle counts. No mass release or particle conversion information. See also Ref. 43. Single point estimate made, but not very useful.

No. 23A This is Vol. 2, Part B. Need Vol. 1 and Vol. 2A. No data on where, what, or how much. Results for trials 10 through 17, but generally uninterpretable without Vol. 1 and Vol. 2, Part A. No estimate made.

No. 25 A theoretic discussion of the analyses that would be applied to the Woodlot series. No data were available at the time of this report, and only the first part of the experimental series had actually been carried out. The number of releases in the first part of the experimental period (Sept. 25 to Dec. 15, 1967) is given, but no information on quantity of release or exposures measured is reported. No estimate made.

No. 33 Only two title pages, three summary pages, and three document pages available. The full volumes 1 and 2 are required for analysis. No estimate made.

No. 43 Published version of Ref. 17 (above).

b. References That Are Not Directly Relevant

No. 14 Only zinc sulfide disseminated. No zinc cadmium sulfide.

No. 15 Continuation of Report 14. Theoretic analysis of previous results.

No. 21 Vol. 3 only. Mathematic model only. No measurements, or any specific trial. Mentions the Jungle Canopy experiment.

No. 26 Matagorda Deposition Trials. Tracer materials were glass beads and fluorescent tagged cork (using 8% Velva Glo). No further information on the fluorescent material is given. The data appendices are missing. A technical report on the same experiment is attached.

No. 38 A 1967 abstract of a paper on the FP tracer technique. No data. References to literature on the technique.

No. 39 The experiments mentioned used sulfur dioxide gas. No ZnCdS.

No. 40 A discussion of washout of particles and gases by rain. A reference is given to a paper describing 20 experiments with FPs.

No. 42 A 1965 paper on the FP technique from Metronics (Leighton, Perkins, Grinnell, and Webster). Describes various products mentioned in the other references: Valron Estersil (blended with ZnCdS pigments at 0.5% to improve flow), which is a hydrophobic silica product manufactured by DuPont (U.S. Patent No. 2,657,149, issued 27 Oct. 1953); and FP 2266, which was made by New Jersey Zinc (NJZ) Company and is now made by U.S. Radium Corp. (USRC). FP 2267 is the same as 2266, except that it has been selected to have a maximum number of particles between 0.75 μm and 3.0 μm in diameter and has been treated to improve flow. Particle-size distributions are given for two lots of 2266: Lot A, mass mean diameter 3 μm, number median diameter 1.8 μm, ppg of 1.7×10^{10}; and Lot B, mass mean diameter 3.2 μm, number median diameter 1.4 μm, ppg of 1.4×10^{10} (but distribution unacceptable for tracer technique).

There is evidence in the paper that experiments up to 10 years old have been re-examined, but no data are presented.

Rotorod efficiencies observed to range from 28% to 73% for FP 2266 with different particle-size distributions. First for MMD of 1.8 μm, 7.9×10^{10} ppg, second for MMD 3.1 μm and 1.6×10^{10} ppg.

Several tables contain particle counts per gram.

Material	Lot	ppg	Uncertainty
NJZ 2266	9BM5	5.94×10^{10}	2.9%
USRC 2267	WS-11	1.33×10^{10}	2.5%
USRC 2267	128	1.56×10^{10}	2.4%
USRC 2267	140	1.56×10^{10}	4.3%
USRC 2267	142	1.38×10^{10}	5.2%
USRC 3206	129	1.62×10^{10}	3.2%
USRC 3206	141	2.24×10^{10}	5.3%
USRC 3206	143	1.64×10^{10}	6.6%

USRC 3206	145	1.33×10^{10}	4.4%
NJZ 2266	8BG505	3.3×10^{10}	
USRC 2267	1339-2	1.7×10^{10}	
USRC 2267	WS-11	0.97×10^{10}	
USRC 2267	12-21	1.2×10^{10}	
USRC 2210	0067	1.35×10^{10}	

Field Experiment	Lot	ppg
138	H324-2	3.3×10^{10}
139	DPG12-21	1.2×10^{10}
140	1339-2	1.6×10^{10}

No. 46 A 1967 revision of a technical manual (originally published 1963, revised 1964) for the FP tracer technique. Describes the assessment of particle collections on membrane filters and impactors (Rotorods).

No. 47 QR SAL 448-4 (July-Sept. 1960) of the Stanford Aerosol Laboratory (SAL), Stanford University. Mention is made of an Operation LAC, with four trials mentioned (A-1, A-2, A-3, B-1), and references are given—apparently those trials were conducted at Dugway, because the sampling data are included in a report entitled "Data Supplement to DPGTR 227." (Dugway Proving Ground Technical Report). There are indications of use of >50 lb of FPs in tower trials at Dugway. Particles-per-gram counts are given for several lots. It is mentioned that the ppg for WINDSOC Lots 7 through 13 are listed in a previous report (QR SAL 448-3). These were obtained through aerosolization tests, the nature of which is not discussed.

Test	FP Material	ppg
34A & B	2266 Lot CEP 8000-0002	4.51×10^{10}
42A & B	2267 Lot 14 WINDSOC	1.34×10^{10}
43A & B	2267 Lot 15 WINDSOC	0.92×10^{10}
44A	2267 Lot 11 WINDSOC	1.34×10^{10}
44B	2267 Lot 11 WINDSOC	1.28×10^{10}
44C	2267 Lot 11 WINDSOC	1.26×10^{10}
45A & B	2210 Lot 0067 Experimental	1.35×10^{10}

Appendix B

The last lot is described as being used extensively by the GE Hanford group. The report mentions a series of small-scale trials (SAL FE 132, 29 Aug. 1960) (FE, field experiment) in which mixtures of FP 2267 and FP 2210 were aerosolized and sampled downwind. These were to be described in a subsequent report. Six releases were completed in an afternoon, involving release rates of 2-3 g/min (period not specified) with samplers 700 ft downwind.

No. 48 June 1955 Operational Manual (reprinted 1958) for the Stanford Fluorescent-particle Technique. Some discussion of the toxicity of ZnCdS. For example,

> "When the aerosol generator shortly to be described is operated near its maximum output, i.e., about 10 g of FP material per minute, the above dosage tolerance figure of 10^{10} particle-minutes per liter could only be reached by an individual who breathed directly in the effluent air stream from the generator for a period of several hours.
>
> "During any reasonable operation of the equipment described in this manual, there is a safety factor of 1,000 to 1,000,000 or more between the dosages which could possibly be encountered by personnel in the neighborhood of the operations and the dosages which have been experimentally demonstrated (with nearly equivalent material) to be absolutely harmless.
>
> "Atmospheric FP tracer experiments may be run hundreds of times over the same populated area without subjecting any inhabitant to more than one millionth of the proven safe dosage.
>
> "The potentially toxic effects of any surface depositions of FP material produced by the experimental operations are nil."

No. 61 A 1970 review by GCA Corp. of meteorologic aspects of a high altitude release of ZnCdS over the West Coast of the United

States. No information on actual releases. Contains a discussion of the potential air pollution and deposition hazards, and some estimates of maximum human exposures to cadmium that could arise, coupled with comparisons that show these exposures are negligible compared with human daily intakes of cadmium.

No. 71 Duplicate of No. 48.

Unnumbered W.D. Crozier and B.K. Seely. Concentration distributions in aerosol plumes three to twenty-two miles from a point source. Trans. Am. Geophys. Union 36 (1955) 42-52. This describes results of dispersion studies in the plains of New Mexico in June and July of 1952, and some work in Australia, using pigment No. 2210 from NJZ. This pigment is described as ZnS, not ZnCdS.

Pigment 2210 is mentioned in Nos. 42, 47, and 27. The first is just a measurement of ppg. The second is a measurement of ppg, together with a small-scale aerosolization test. The third is a set of experiments on dispersion into a forested area.

c. References That Are Not Available for This Report

No. 4(e) Dugway Proving Ground, 7 Apr. 1954.
No. 5 Dugway Proving Ground, Oct. 1954.
No. 6 Evaluation of the Cluster B133 Filled BG (12 pp.) "Operation Polka Dot"
No. 7 Dugway Proving Ground.
No. 8 DEW II. An experimental study of long range aerosol cloud travel involving ground deposition of biologic spore material (113 pp.).
No. 9 Minneapolis, Minn., St. Louis, Mo. 15 Jan.-24 Mar. 1953. Behavior of aerosol clouds within cities (82 pp.). Almost certainly refers to the same studies as Ref. 35.

No. 10 Rosemont, Minn. Sept.-Oct. 1953. Preliminary field trials of the aerosol X1A. Abn dry agent dissemination unit (78 pp.).
No. 11 San Francisco Bay, Redwood City, Calif. 21 and 26 Mar. 1956.
No. 12 Continental U.S. East of Rocky Mountains. 30 Nov. 1957, 6 Feb. 1958, 25 Apr. 1958, 20 Mar. 1958.
No. 13 WINDSOC. Most of the documentation is missing.
No. 23 Victoria Diffusion Trials, Vol. 1 (and also Vol. 2, Part A).
No. 34 Operation Moby Dick, San Francisco Bay, Calif. Sept. 1950.
No. 44 A repeat of the first volume of Ref. 19.
No. 45 Preliminary air pollution survey of cadmium and its compounds, Y.C. Athanassiadis, Contract PH 22-68-25, Oct. 1969, U.S. Department of Health, Education, and Welfare. 235 pp.

D. A Chronology of the Earliest Experiments

Stanford University (Grinnell, Perkins, Webster, later Hutchison)

Jan. 1946	Contract W-18-035-CWS-1256	Original 2-year contract, running 1/1/46 through 12/31/47
June 1947	Contract W-18-035-CM-147	Next 2-year contract
Dec. 1947	Contract W-18-035-CWS-1256	Original contract extended to 2/28/48 (Ref. 36 covers the results of Contract W-18-035-CWS-1256 through the end of 1947.)
Feb. 1948	Contract W-18-035-CM-147	Both contracts combined to run under this number until 12/31/49

The results obtained in W-18-035-CM-147 are not documented in any of the available reports. From subsequent experiment numbering and discussions, it appears that up to 14 experiments were performed over populated areas (numbered FE 1 through FE 14), probably in Palo Alto.

In Ref. 37, the following bimonthly reports are referred to: BMR 4, Oct.-

Nov. 1948; BMR 9; BMR 11, Apr.-May 1949; BMR 13; BMR 14, Oct.-Nov.-Dec. 1949.

Ref. 37 also mentions that experiments FE 13 and FE 14 showed the feasibility of using FPs in 1-lb quantities for a dispersion over a distance of 4 miles or more.

Feb. 1950   Contract DA-18-108-CML-450   Continuation of studies under W-18-035-CM-147.

These are documented in Ref. 37 quarterly reports. However, QR 4 is missing, and any quarterly reports beyond QR 7 are also missing (including any final report).

E. CHRONOLOGY AND CROSS-REFERENCING OF THE EXPERIMENTS

This chronology incorporates many of the loose ends that were noticed but is almost certainly not complete. It was compiled by noting in each reference any cross-references to other experiments. All dates are said to be those of FP releases described in the cross-reference, or of documents that are said to describe FP releases in the cross-reference. In some cases, there are guesses based on nomenclature, indicated by a ? under the reference.

The notation ?(5) indicates a cross-reference to some release, and it is possible that the unavailable reference given in the parentheses corresponds to that cross-reference. It is also possible that these two are distinct.

The list below shows the following: (1) a range of dates during which releases were said to take place, (2) the reference-list number (*Ref.*) in which that set of releases is described in detail, or (3) the reference-list number (*XRef.*) in which that release is mentioned but is not described in detail. A reference number in parentheses indicates that the reference is in the list but was not in the set of references available for this report.

Appendix B

The unnumbered reference is indicated with an asterisk (*).

Note that the people originally at Stanford subsequently or concurrently founded or worked at or for Metronics.

Dates	Ref.	XRef.	Comments
1/46-12/47	36		Stanford: Contract W-18-035-CWS-1256. Original experiments at SAL. One release 10/47.
6/47-?/49		37	Stanford: Contract W-18-035-CM-147. Possible field experiments FE 1 through FE 12. Certainly field experiments FE 13 and FE 14 involving pound quantities.
2/50-?	37		Stanford: Contract DA-18-108-CML-450. Field experiments FE 18 through FE 30. Last available QR is Oct. 1951.
1950		2	Dugway. Summer. FP, BG, and AP1 generated simultaneously.
10/50		2	Dugway. Special Report 142. San Francisco.
1951		2	Dugway test. FP with AB1 and bacterium *Tularense* (more than four trials)
3/52-4/52	8		DEW I. N. and S. Carolina, and Georgia.
1952	(8)	2	DEW II. N. and S. Carolina, and Georgia
6/52-7/52		*	New Mexico (FP 2210, possibly not ZnCdS).
1/53-10/53	35		Minneapolis, St. Louis, and Winnipeg. Aerosol clouds in cities. Contracts DA-18-064-CML-1856 (Stanford) and DA-18-064-CML-2282 (Parsons).
2/53-3/53	30		SELTZER (Dugway), and WHITEHORSE (Florida).
5/53-6/53	3		Dugway tests BW 6-52 and BW 5-52.
9/53-10/53	(10)		Rosemont, Minn.
1953		2, 41	Stanford. Contract DA-42-007-403-CML-111.
1953		2	Stanford. Contract DA-18-108-CML-450: "Dispersion of aerosols and travel of aerosol clouds." This contract number is the same as given above (see 2/50-?), so it might be the final report for the contract.
1953		2	Stanford. Contract DA-18-064-CML-1856.

			"Research concerning the propagation of airborne agents for work over cities."
1/54-3/54	4		Dugway tests BW 8A-1-54 through BW 8A-5-54, omitting 8A-3-54.
1/54-3/54	?(4e)	4	Dugway test BW 8A-3-54; suggested by nomenclature of XRef.
1954	?4	2	Dugway. "A number of trials producing aerosols of BG and FP."
1954	(34)	2	MOBY DICK, Calif.
1954/55	?(5)	6	Trials BW 8B-1-54 through BW 8B-6-54 suggested by numbering.
5/55	6		TROUBLE MAKER. Dugway trials BW 8B-7-54 and BW 8B-8-54.
7/55	(6)		POLKA DOT.
8/55-11/55	29		GOOF. Dugway. Trials BW 1A-1-56 through BW 1A-5-56.
1955	2		Camp Cooke, Calif.
1955	?	2	Unexplained reference to "St. Jo munitions expenditure panel" might indicate a named operation, possibly the Camp Cooke operation.
3/56	(11)		San Francisco Bay, Redwood City, Calif.
1956	(7)		Dugway Proving Ground.
11/57-3/58	(12)		Continental U.S., east of Rocky Mountains. 30 Nov. 1957, 6 Feb. 1958, 25 Apr. 1958, 20 Mar. 1958.
4/58	31		Dugway. BW 398-A Trials 8, 9, 10, 11.

Appendix B

FP use appears to be routine and hardly commented on at Dugway by this time.

Dates	Ref.	XRef.	Comments
8/59-2/60	13		WINDSOC. Contract DA-42-007-403-CMl-432.
before 9/60		47	LAC. Dugway (?). Four trials mentioned (A-1, A-2, A-3, B-1). "Data Supplement to DPGTR 227" is said to contain the FP sampling data. Also, Stanford Semiannual Report SALR 111-19, June 1959 is Part I, and Technical Report No. 86, Analysis of LAC-58 Trials B-1 and A-3, Sept. 1960, is Part II of a report on Operation LAC.
before 9/60		47	Forest canopy trials (Test plan 486).
8/60		47	Mention of field experiment. SAL FE 132. At least six releases.
?		41	Dugway Trial 507 B-7 is said to have used an FP discussed in Ref. 41 (presumably occurs before Trial 508—see 1/61 below).
1/61		18, 47	Dugway. DPGTM 1045, Field calibration of the L 23 FP Disseminator, Bio 508A. At least five releases. Ref. 47 mentions "BW Trial 508-3" and "Trial 508-9," and indicates the possible use of >50 lb FP in this trial.
4/61-8/61	33		Dallas Tower Studies. Texas. Contract DA-42-007-CML-504.
8/61-1/63		19, 21	Bendix Corp. Bendix Systems Division. Jungle Canopy Penetration. Vol. 1. Diffusion Measurements. Contract DA42-007-530. Final Report (Aug. 1961-Jan. 1963) (by Hamilton et. al., according to p. 1-32 of text of Ref. 19).
1/62-11/62	41		Palo Alto. Contract DA 42-007-CML-543. Field experiments FE 134 through FE 140. Metronics FP efficiency test of Rotorods.
6/62-11/62	16		Oklahoma, Texas, Washington, Nevada.
7/62		18	Dugway. DPGTM 1052. Calibration of the

			Model D-1 dry particulate disseminator designed for the L-20 aircraft, Bio 508B. July 1962. At least three releases.
5/63-8/63	18		Dugway.
5/63-3/65	17, 43		St. Louis
8/63		18	Dugway. DPGTM 1058, Drone Delivery System AN/USD-2 (XAE-3), Phase III, biologic studies. Aug. 1963. At least five releases.
11/63		19	Meteorology Research, Inc. Report 63-FR-108. Dissemination and Evaluation of a Tracer Material Release (Big Jack) (U), Vol. 1, by T.B. Smith and F. Vukovich. Contract DA42-007-AMC-15(X). Final Report (29 Nov. 1963). AD 347, 791. CONFIDENTIAL. (Text of Ref. 19, p. 1-31 indicates authors as Smith and Leavengood).
1/64-8/64	19		Chippewa Forest, Minn. Deciduous Forest Diffusion Study. Contract DA-42-007-AMC-48(R).
6/64-8/64		21	Diffusion under a Jungle Canopy (Ref. 21 is Vol. 3, theoretic analysis only. Vols. 1 and 2 are not available.)
2/64-2/66	22		Fort Wayne. Urban Diffusion Project. Contract DA-42-007-AMC-37(R).
3/64		19	Geophysics Corp. of America. Technical Report 64-3-G. Meteorological Prediction Techniques and Data System, by H.E. Cramer et. al. Contract DA42-007-CML-552. Final Report. AD444, 197.
3/64		20	Metronics. Field experiments FE 146 mentioned.
1/64-2/65	1		Central Alaska.
6/65-6/66	(23)		Victoria, Tex. Diffusion trials. July-Aug. 1965, 9-29 July 1966.
10/65-5/67	20		San Francisco. Contract DA-42-007-AMC-2-40(R). From Metronics. Field experiments FE 151 through 157.
1966		19	I.A. Smith and M.E. Singer. Personal conversa-

APPENDIX B

		tions. (Text of Ref. 19 indicates experiments at Brookhaven National Laboratory.)
6/67-7/67	24	Oceanside, Calif. Oceanside diffusion shoreline project. Contract DA 42-007-AMC-180(R)
8/67	24	T.B. Smith, and K.M. Beesmer. Bolsa Island Meteorological Investigation. MRI Report FR-650 for Bechtel Corp., 58 pp., 1967. Eight releases mentioned.
9/67-12/67	25	WOODLOT.
6/68	32	Dugway Proving Ground. "T3-665" trials testing the dissemination efficiency of the Mark IX disseminator. Final Report. Technology test of FP fluidizers, by W.A. Brown and J.E. Frese, RDT&E Project 1V025001A128, June 1968.
10/68-9/69	27	Pack Forest, Wash. Forest diffusion. Grant DA-AMC-28-043-68-G8.
8/69-10/69	28	MATE. First field test. Contract DA 42-007-AMC-339(Y)
2/70-3/70	32	Dugway. Check test of West Vertical Grid.

F. COMMENTS ON CHRONOLOGY AND CROSS-REFERENCING

The series of field experiments named FE 1 upwards was started by Stanford, and apparently continued by Metronics. Note that there are gaps in the numbering in the documents available.

Ref. 32 used a Mark IX disseminator and refers to the "T3-665" trials testing the efficiency of this disseminator. However, there is little, if any, reference to earlier "Mark" disseminators. Other references are to the D-1 (L-20) and the L-23 disseminators or to a Stanford generator or a Skil blower.

4. REFERENCE LIST

1. T.B. Smith and K.M. Beesmer. 1967. Dissemination and Evaluation of

a Tracer Material Release. West Site (U) Vol. 2—Part B. Report to Deseret Test Center, Fort Douglas, Utah, from Meteorology Research, 464 West Woodbury Road, Altadena, Calif. Project DESERET. Contract DA 42-007-AMC-162(Y) Phase II. MRI Document MR165 FR-288. Mar. 1967.
2. Special Report 273. Comparison of Simulant Decay Rates in Field Tests (U). Program Research, Field Operations and Meteorology Branches, Assessment Division, Fort Detrick, Frederick, Md., Nov. 1956 (supersedes Assessment Division Test Report A-294).
3. Comparative Diffusion of Dissimilar Agents. PGR 149 W5, 6-52. Downwind travel of simulant agents released simultaneously, BW 5 and 6-52. Dugway Proving Ground Report 149. Projects 4-98-05-005 and 4-98-01-001. 5 Feb. 1954.
4. Trial Reports from Dugway Proving Ground, Relationship of Dosages to Source Strength for BG and an Inert Tracer. 24 Feb. 1954.
    Report DPG, BW 8A-1-54, 24 Feb. 1954
    Report DPG, BW 8A-2-54, 9 Mar. 1954
    Report DPG, BW 8A-4-54, 22 Mar. 1954
    Report DPG, BW 8A-5-54, 12 Apr. 1954
5. Dugway Proving Ground, Utah. Oct. 1954 (not available for this report).
6. Trial Report from Dugway Proving Ground. BWAL, BW 8B-7 and 8-54, Operation "Trouble Maker." Relationship of dosages to source strength for BG and an inert tracer. BWALTR 27. 18 July 1955.
7. Dugway Proving Ground, Utah. 1956 (not available for this report).
8. An Experimental Study of Long Range Aerosol Cloud Travel. Special Report 162. U.S. Army Chemical Corps Biological Laboratories, Camp Detrick, Frederick, Md., 1 Aug. 1952.
9. Behavior of Aerosol Clouds Within Cities (Classified) (82 pp.). Minneapolis, Minn., 15 Jan.-24 Mar., St. Louis, Mo., 1953 (not available for this report).
10. Preliminary Field Trials of the Aerosol X1A. (Classified) (78 pp.) Abn dry agent dissemination unit. Rosemont, Minn., Sept.-Oct. 1953 (not available for this report).
11. San Francisco Bay, 21 and 26 Mar. 1956, Redwood City, Calif. (not available for this report).

APPENDIX B 153

12. Continental U.S., 30 Nov. 1957, East of Rocky Mountains, 6 Feb. 1958, 25 Apr. 1958, 20 Mar. 1958 (not available for this report).
13. T.B. Smith and M.A. Wolf. Intermediate Scale Particulate Cloud Travel—Project WINDSOC—Part 1—Technical Report. Meteorology Research, 2420 N. Lake Ave., Altadena, Calif. MRI Report 60-23. Prepared for U.S. Army Chemical Corps Proving Ground under Contract DA-42-007-403-CMI-432. July 1960.
14. D.A. Haugen and J.J. Puquay, eds. The OCEAN BREEZE and DRY GULCH Diffusion Programs. Vol. 1. Research Report. Meteorology Laboratory, Air Force Cambridge Research Laboratories, Office of Aerospace Research, U.S. Air Force. AFCRL-63-719(I). Hanford Doc. HW-78435. Nov. 1963.
15. D.A. Haugen and J.H. Taylor, eds. The OCEAN BREEZE and DRY GULCH Diffusion Programs. Vol. 2. Research Report. Meteorology Laboratory, Air Force Cambridge Research Laboratories, Office of Aerospace Research, U.S. Air Force. AFCRL-63-719(II). Dec. 1963.
16. T.B. Smith and M.A. Wolf. Vertical Diffusion from an Elevated Line Source over a Variety of Terrains. Part A. Final Report to Dugway Proving Ground. Meteorology Research, 2420 North Lake Ave., Altadena, Calif. Contract DA-42-007-CML-545. Mar. 31, 1963.
17. F. Pooler. A Tracer Study of Dispersion over a City. Paper 66-28 at the 59th Annual Meeting, Air Pollution Control Association, San Francisco, Calif., June 20-24, 1966 (from the Air Resources Field Research Office, ESSA, at the Robert A. Taft Sanitary Engineering Center, Division of Air Pollution, U.S. Department of Health, Education, and Welfare, Cincinnati, Ohio).
18. Calibration of the Model D-1 Dry Particulate Disseminator with Green Fluorescing FP, DPGTP 508C. USATECOM Project 5-3-9030-10. DPGTM 1060. Case File 3164. Biological Branch, Test Design and Analysis Division, Technical Plans and Evaluation Directorate, U.S. Army Test and Evaluation Command, Dugway Proving Ground, Utah. Nov. 1963.
19. Deciduous Forest Diffusion Study, Report 3004, Final Report. Applied Science Division, Litton Systems, Minneapolis, Minn. Contract DA-42-007-AMC-48(R). Project 1T062111A128, for the

Meteorology Division, Deseret Test Center, Building 103 Soldiers Circle, Fort Dugway, Utah. June 1969.

Vol. 1: M.H. Tourin and W.C. Shen. Diffusion studies in a deciduous forest.

Vol. 2: M.H. Tourin, R.J. Rickett, and W.C. Shen. Detailed program description.

Vol. 3: M.H. Tourin and W.C. Shen. Detailed technical data supplement.

20. J.A. Murray, T.S. Brown, and F.X. Webster. FP Tracer Co-Dispersal and Sampling Studies. Technical Report 135. Metronics Associates, 3201 Porter Drive, Stanford Industrial Park, Palo Alto, Calif. Contract DA-42-007-AMC-240(R). Task IV025001A128, Meteorological Aspects of CB Program. U.S. Army Dugway Proving Ground, Utah. Dec. 1968.

21. Diffusion under a Jungle Canopy. Final Report. Vol. 3: Mathematical Model. Meteorological Research Laboratory, Melpar, 7700 Arlington Blvd., Falls Church, Va. Contract DA 42-007-AMC-33(R) for U.S. Army Dugway Proving Ground, Dugway, Utah. Oct. 1967.

22. G.R. Hilst and N.E. Bowne. A Study of the Diffusion of Aerosols Released from Aerial Line Sources Upwind of an Urban Complex. Vol. 1, Final Report; Vol. 2, Data Supplement. Travellers Research Center, 250 Constitution Plaza, Hartford, Conn. Contract DA-42-007-AMC-37(R) under RDT&E Project 1V025001A128 Meteorological Aspects of CB Program, for U.S. Army Dugway Proving Ground, Salt Lake City, Utah. July 1966.

23. R.L. Miller, ed. Victoria Diffusion Trials. MR166 FR-374. Meteorology Research, 464 West Woodbury Road, Altadena, Calif. Contract DA 18-064-AMC-422(A) for Fort Detrick, Frederick, Md. May 1966.

Vol. 1: Final Report (not available for this report)

Vol. 2: Parts A and B (not available for this report)

24. T.B. Smith, and B.L. Niemann. Shoreline Diffusion Program, Oceanside, Calif. Report MRI 169 FR-860. Meteorology Research, 464 W. Woodbury Road, Altadena, Calif. Contract DA 42-007-

AMC-180(R) for Deseret Test Center, Fort Douglas, Utah. Nov. 1969.
25. G.R. Hilst and G.T. Csanady. Aerosol Diffusion over Woodlot Complexes. Theoretical and Empirical Assessments of Atmospheric Dispersion and Deposition Processes over Inhomogeneous Terrain (Woodlots). Phase I Report. DAAD09-67-C-0100(R). Travellers Research Center, 250 Constitution Plaza, Hartford, Conn. RDT&E Project IV025001A128 Meteorological Aspects of CB Program for U.S. Army Dugway Proving Ground, Dugway, Utah. Aug. 1968.
26. G.P Ettenheim, Jr. and C.L. Crum. MATAGORDA Deposition Trials. Field Report. MR167 FR-468. Meteorology Research, 464 West Woodbury Road, Altadena, Calif. Contract DA 18-064-AMC-422(A) for Fort Detrick, Frederick, Md. 31 Jan. 1967.
27. Dispersion of Air Tracers into and Within a Forested Area. College of Forest Resources, University of Washington, Seattle, for U.S. Army Electronics Command, Atmospheric Sciences Laboratory, Fort Huachuca, Ariz. Grant DA-AMC-28-043-68-G8. Technical Report ECOM-68-G8-1. Sept. 1969.
  Vol. 1: L.J. Fritschen, C.H. Driver, C. Avery, J. Buffo, R. Edmonds, and R. Kinerson (objectives, methods, site description, preliminary data).
  Vol. 2: L.J. Fritschen, C.H. Driver, C. Avery, J. Buffo, R. Edmonds, R. Kinerson, and P Schless (tabulated data and dispersion patterns).
  Vol. 3: L.J. Fritschen, C.H. Driver, C. Avery, J. Buffo, R. Edmonds, R. Kinerson, and P Schless (analysis and interpretation).
28. J.K. Allison, A.V. Duffield, and J.M. Morton. Meteorological Analog Test and Evaluation: First Field Test Operation. Final Report. Meteorological Research Laboratory, Melpar Division of American Standard, 7700 Arlington Blvd., Falls Church, Va. Contract DA 42-007-AMC-339(Y), Task VIII B, for U.S. Army Deseret Test Center, Fort Douglas, Utah. June 1970.
29. Comparison of Decay Rates for *Bacillus globigii* and Zinc Cadmium Sulfide. BW 1A-56, Operation "GOOF." DPGR 175. Dugway Prov-

ing Ground Report, BW Assessment Directorate, Project Order 0016. 27 Apr. 1956; Trial Report, BWAL, BW 1A-1, 2, and 3-56. Comparison of Decay Rates for BG and FP. Operation "GOOF." BWALTR 38. 7 Nov. 1955; Trial Report, BWAL, BW 1A-4 and 5-56. Operation "GOOF." CMLRE-DU-MBW. BWALTR 40. 21 Dec. 1955.

30. An Experimental Investigation of Viable Aerosol Travel from Sea to Land. Special Report 193. U.S. Army Chemical Corps Biological Laboratories, Camp Detrick, Frederick, Md. 24 Sept. 1953.

31. Trial Record 230, BW 398-A, Trials 8, 9, 10, and 11. Test Design and Analysis Office, Technical Operations Directorate, U.S. Army Chemical Corps Research and Development Command, U.S. Army Chemical Corps Proving Ground, Dugway, Utah. May, 1958.

32. A.T. Hereim and J.E. Frese. Check Test of West Vertical Grid, Dugway Proving Ground, Utah. Summary Report. RDTE Project 1-X-6-65704-D-634-06. USATECOM Project 5-CO-413-000-013. DTC Project DTC B-008. Deseret Test Center, Fort Douglas, Utah. Oct. 1970.

33. P.B. MacCready, T.B. Smith, and M.A. Wolf. Vertical diffusion from a low altitude line source—Dallas Tower Studies, Vol. 1. Final Report MR161 FR-33 from Meteorology Research, 2420 North Lake Ave., Altadena, Calif., for the U.S. Army Chemical Corps, Dugway Proving Ground. Contract DA-42-007-CML-504. Dec. 1961. (Note: only two title pages and pp. I-iii and 1-3 were available for this report.)

34. Operation Moby Dick. Sept. 1950. San Francisco Bay, Calif. (not available for this report).

35. Behavior of Aerosol Clouds within Cities. Joint quarterly reports submitted by Stanford University and the Ralph M. Parsons Company to the U.S. Army Chemical Corps. Contract Nos. DA-18-064-CML-1856 (Stanford) and DA-18-064-CML-2282 (Parsons).
    A. JQR No. 1, July-Sept. 1952
    B. JQR No. 2, Oct.-Dec. 1952
    C. JQR No. 3, Jan.-Mar. 1953
    D. JQR No. 4, Apr.-June 1953 (missing some pages)
    E. JQR No. 5, July-Sept. 1953 (missing some pages)

F. JQR No. 6, Oct.-Dec. 1953, Vols. 1 and 2 (Vol. 2 is missing some pages)
36. S.W. Grinnell, W.A. Perkins, F.X. Webster. Bimonthly Report 11 submitted by Stanford University to the U.S. Army Chemical Warfare Service Research and Development Program. Contract W-18-035-CWS-1256, Sept.-Oct. 1947. (BMR 1 through 13 and the final report are all available. Only BMR 11 describes any field release. The others imply small (milligram quantity) releases into an air chamber, and "field work of a minor nature.")
37. Quarterly reports submitted by Stanford University to the U.S. Army Chemical Corps Research and Development Program. Contract DA-18-108-CML-450.
   A. QR 1, Feb.-Mar. 1950 (tests at Palo Alto)
   B. QR 2, May-June 1950 (tests at Dugway Proving Ground)
   C. QR 3 Aug.-Sept.-Oct. 1950 (San Francisco and adjacent area)
   D. QR 4, Missing
   E. QR 5, Feb.-Mar.-Apr. 1951 (measurement of urban temperature gradients)
   F. QR 6, May-June-July 1951 (meteorologic studies)
   G. QR 7, Aug.-Sept.-Oct. 1951 (meteorologic studies; FP size measurement)
38. F.X. Webster. The FP Atmospheric Tracer Technique. Metronics Associates, Stanford Industrial Park, Palo Alto, Calif. (Abstract of paper presented at Conference on Air Pollution in California, held Oct. 30, 1967 at San Jose State College, Calif.) (see also Ref. 42).
39. H.E. Cramer, F.A. Record, and H.C. Vaughan. The Study of the Diffusion of Gases or Aerosols in the Lower Atmosphere. AFCRC-TR-58-239. ASTIA Doc. 152582. Massachusetts Institute of Technology, Department of Meteorology. Final Report. Contract AF 19(604)-1058. Sponsored by the Geophysics Research Directorate of the Air Force Cambridge Center, Air Research and Development Command. 15 May 1958.
40. L.M. Vaughan and W.A. Perkins. The Washout of Aerosol Particles and Gases by Rain. Technical Report 88, Aerosol Laboratory, Stan-

ford University, Stanford, Calif. Contract DA-42-007-403-CML-448. U.S. Army Chemical Corps Research and Development Program. Jan. 1961.
41. F.X. Webster. Collection Efficiency of the Rotorod FP Sampler. Technical Report TR 98. Aerosol Laboratory, Metronics Associates, 3201 Porter Drive, Palo Alto, Calif. Contract DA 42-007-CML-543 for the U.S. Army Chemical Corps Research and Development Program. 31 Jan. 1963.
42. P.A. Leighton, W.A. Perkins, S.W. Grinnell, and F.X. Webster. 1965. The fluorescent particle atmospheric tracer. J. Appl. Meteorol. 3:334-348.
43. F. Pooler, Jr. 1966. A tracer study of dispersion over a city. J. Air Pollut. Control Assoc. 16:627-631.
44. M.H. Tourin and W.C. Shen. Diffusion Studies in a Deciduous Forest. Vol. 1, Final Report. Report 3004, AD 856703L. June 1969. 325 pp. (part of Ref. 19).
45. Y.C. Athanassiadis. Preliminary Air Pollution Survey of Cadmium and Its Compounds. Contract PH 22-68-25. U.S. Department of Health, Education, and Welfare. Oct. 1969. 325 pp. (not available for this report).
46. FP Tracer Counting Manual. Technical Manual 163-2. Aerosol Laboratory, Metronics Associates, Stanford Industrial Park, Palo Alto, Calif. Prepared for U.S. Weather Bureau, Dept. of Commerce, Contract CWB-10635. 17 May 1963 (rev., 11 Aug. 1964).
47. QR No. 448-4. July-Sept. 1960. Aerosol Laboratory, Stanford University, Stanford, Calif. Contract DA-48-007-403-CML-448. U.S. Army Chemical Corps Research and Development.
48. P.A. Leighton, director. The Stanford Fluorescent-particle Tracer Technique—An Operational Manual. Dept. of Chemistry, Stanford University. June 1955 (second printing, June 1958).
61. H.E. Cramer, R.N. Swanson, and A.G. Tingle. Review of the Meteorological Aspects of a High Altitude Release. GCA Corp., GCA Technology Division, Bedford, Mass., for U.S. Army Deseret Test Center, Fort Douglas, Utah. 9 Sept. 1970.
71. Duplicate of Ref. 48.
Unnumbered. W.D. Crozier and B.K. Seely. 1955. Concentration distri-

butions in aerosol plumes three to twenty-two miles from a point source. Trans. Am. Geophys. Union 36:42-53.

DESCRIPTION OF TEST

Name of test: *Comparison of Simulant Decay Rates in Field Tests*

Reference-list number: 2

Reference: Special Report 273. Comparison of Simulant Decay Rates in Field Tests (U). Program Research, Field Operations and Meteorology Branches, Assessment Division, Fort Detrick, Frederick, Md. Nov. 1956 (supersedes Assessment Division Test Report A-294).

Principal object:
To obtain a definitive estimate of the downwind physical decay of aerosols of BG and FP and their relative decay rates; to compare estimates of these relationships obtained from three kinds of dissemination; to relate any differences between estimates to weather factors; to compare these estimates with those obtained elsewhere.

Site selection:
"Of the many sites considered for this test, Camp Cooke, California, offered the best combination of characteristics for efficient field testing." The principal factor was climate, but other factors included terrain, military control of the land, and operational factors. The prevailing wind from the northwest during summer ensured that trials could be run on consecutive days shortly after sunset.

Number and nomenclature of releases:
Three phases (A, B, and C) with different release mechanisms (for FP or BG) in each phase. Thirty-nine trials were conducted, numbered 1 through 39, with 1-20, 26-27 in Phase A, 21-25 in Phase B, and 28-39 in Phase C. Difficulties with the sampling of BG led to a complete dis-

carding of data from trials 1-4 and 9-12, with trials 5-8 considered somewhat unreliable for BG.

Test conditions:
Phase A: Two E61R4 (bomblets) filled with 40 g each of FP, and located 40 ft apart, were fired simultaneously.
Phase B: A single Stanford generator (80 g hopper) disseminated at approximately 8 g/min for 4 min at ground level.
Phase C: As Phase B, but at 13 ft above ground.

Samplers were placed on a 90-degree circular sector grid at 100, 200, 400, 600, 1,200, 1,760, and 2,640 yd, with 20, 40, 60, 80, 80, 80, 120 stations at each radius, respectively. The samplers were Millipore type with a flow rate of 12.5 L/min.

Test material:
ZnCdS from New Jersey Zinc Company, CEP 8000, Lot 0002 with 2% magnesium silicate. No estimates of particles per gram or dissemination efficiencies are given in the report, but this lot was tested as having 4.51×10^{10} ppg in Ref. 47 (QR SAL 448-4; see Section 4, "Miscellaneous Notes."

Place of release:
Camp Cooke, California, in scrub-brush desert, 2.5 miles downwind from the ocean.

Dates, times, and quantities of release:
No information is given on the dates or times of releases. All that can be inferred is that they were some time in 1955. No more accurate information is given on release quantities than presented in "Test conditions:" above. With these values, each trial in Phase A (22 trials) released 80 g of FPs, and each trial in Phases B and C (17 trials) released 32 g—a total release of 2304 g.

Appendix B

Communities affected:
Unknown. The location of Camp Cooke was not provided.

Distances from releases to affected communities:
Greater than the maximum radius of the sampling grid, 1.5 miles.

Weather conditions:
A neutral temperature gradient prevailed consistently during the hours after sunset.

Other materials released:
There was simultaneous release of BG.

Maximum time-integrated concentrations and concentrations:
The report was concerned with estimating decay curves for FP and BG. No measured counts, concentrations or exposures were presented. Cross-wind-integrated-exposures were tabulated at each sampling radius, and cloud widths (distances between 1/10 maximum exposures) were given. These two measurements have been used to estimate the following maximum exposures and concentrations, by assuming gaussian shaped plumes and (for the instantaneous releases) equal dispersion along and across the plume. Dissemination efficiency has been assumed to be 50% for the bombs, and 100% for the Stanford generators.

The highest measured exposures and estimated concentrations occur at the 100 yd sampling line, and these are the values given for "Unpopulated areas." The upper bound for the "Populated area" was obtained from the estimates for the sampling line at greatest distance from which estimates were available, generally 1.5 miles from the dissemination point for Phase A, and 1 mile from the sampling point for Phases B and C.

Trial	Unpopulated Areas		Populated Areas (Upper Bound)	
	Max Exp., μg-min/m^3	Max Conc., μg/m^3	Max Exp., μg-min/m^3	Max Conc., μg/m^3
		Phase A		
13	7283	55,651	5.9	5.3
14	2260	16,054	9.1	4.5
15				
16	4670	35,683	8.2	6.9
17	2876	23,833	0.6	1.0
18	7191	49,658		
19	1859	15,172	0.8	0.9
20	3984	42,526	8.5	7.3
26	1150	4,516	6.1	1.7
27	3337	22,986	9.7	1.9
		Phase B		
21	1502	369	3.4	0.7
22	2048	507	5.4	1.1
23	608	148	1.0	0.2
24	1351	331	2.4	0.5
25	925	225	1.0	0.1
		Phase C		
28	853	210	12.9	2.8
29	870	213	8.2	1.7
30	741	182	7.6	1.7
31	1023	251	6.5	1.5
32	924	227	5.4	1.1
33	720	174	23.0	4.7
34	1147	275	3.7	0.8
35	1219	297	3.6	0.7
36	666	163	2.5	0.5
37	832	206	10.0	2.2
38	805	198	6.8	1.5
39	497	120	1.6	0.3
Cumulative	60,136	55,651	173	7

The cumulative estimates have been estimated by summing the individual exposures—the wind direction was always similar (from the northwest)—and adjusting the total for Phase A by assuming that the average of the missing trial data was similar to the average of the available trial data. As usual, the "cumulative" maximum concentrations are simply the largest estimated concentrations—the trials were assumed to occur at distinct times. The large difference in concentrations between Phase A and Phases B and C is due to the different dissemination mechanisms—an instantaneous release versus a continuous release.

Other comments:
This report supersedes Assessment Division Test Report A-294. That earlier report might contain the missing raw data.

There are general references to FP use:
- "Dispersion of Aerosols and Travel of Aerosol Clouds." Contract DA-18-108-CML-450, Chemistry Department, Stanford University, 1953.
- "Research Concerning the Propagation of Airborne Agents for Work over Cities." Contract DA-18-064-CML-1856, Chemistry Department, Stanford University, 1953.

These describe experiments over Palo Alto and San Francisco, and so probably duplicate Ref. 37.

There are references to other operations involving FP and BG:
- Summer of 1950, FP, BG, and AP1 generated simultaneously at Dugway.
- 1951, at Dugway, FP with AB1 and bacterium *Tularense* (more than four trials).
- 1954, at Dugway, "A Number of Trials Producing Aerosols of Bg and FP."
- Operation MOBY DICK, "conducted in California in 1954." Special Report 237, "Moby Dick: Sea-to-Land Travel of Simulant Aerosols Generated by the XB-14B Mine (C)." Assessment Division, Camp Detrick, Frederick, Md. Aug. 1955. This is listed as Ref. 34, except that it is said to have occurred in Sept. 1950.
- The operation off Georgia, North Carolina, and South Carolina is

referred to as Operation DEW I. The reference given is the document in Ref. 8 (Dugway Special Report 162).
- An operation named Operation DEW II, involving release of FP and *Lycopodium* spores from an aircraft. This is Dugway Special Report 179 ("An Experimental Study of Long Range Aerosol Cloud Travel Involving Ground Deposition of Biological Spore Material," F&MR Division, Camp Detrick, Frederick, Md., 1 June 1953) and appears to be the second (classified) document in Ref. 8 (not available).
- The Minneapolis, St. Louis, and Winnipeg trials are referred to as "Research Studies of Aerosol Cloud Travel over Cities." Contract DA-18-064-CML-2282, the Ralph M. Parsons Company. This appears to be the study examined in Ref. 35.
- Other referenced trials involving FP are
  Special Report 142. BW trials at San Francisco, Calif., Sept. 1950, PD Division, Camp Detrick, Frederick, Md., 22 Jan. 1951. "Studies in Aerosol Cloud Behavior." Contract DA-42-007-403-CML-111, Chemistry Department, Stanford University, 1953.

There is a reference (p. 3) to the "St. Jo Munitions Expenditure Panel" that might indicate a named operation.

DESCRIPTION OF TEST

Name of test: *Comparative Diffusion of Dissimilar Agents*

Reference-list number: 3

Reference: Comparative Diffusion of Dissimilar Agents. PGR 149 W5, 6-52. Downwind travel of simulant agents released simultaneously, BW 5 and 6-52. Dugway Proving Ground Report 149. Projects 4-98-05-005 and 4-98-01-001. 5 Feb. 1954.

Principal object:
To compare the distribution and downwind-travel characteristics of SM

(*Serratia marcescens*), C2 (*Aspergillus fumigatus*), and FP (2266) released simultaneously; to compare the viability and decay rates of the SM and C2 aerosols; to compare the distribution and downwind-travel characteristics of aerosols of SM, BG, and FP released simultaneously; and to compare the viability and decay rates of the SM and BG aerosols.

Site selection:
Dugway Proving Ground.

Number and nomenclature of releases:
Two releases: BW 5-52 and BW 6-52.

Test conditions:
A portion of the Crops Grid No. 2 was used. The grid sector used lies along an axis of 135° and is 13 miles long by 3.5 miles wide, with grid laterals placed at 0.5-mile intervals. It is bounded to the southwest by Simpson Buttes and Camel Back Mountain, but is flat in the sampling area. Four generators were used to simultaneously disseminate the FP in 90 to 105 s. Filter-type samplers were used.

Test material:
ZnCdS FP type 2266, with 1×10^{11} ppg.

Place of release:
Both tests: approximately 6 miles east-northeast of Simpson Buttes.

Dates, times, and quantities of release:

Test	Date	Start Time	Length, min	Amount, kg
BW 6-52	5/4/53	22:29	1.5	3.534
BW 5-52	6/3/53	00:50	1.75	2.0
Total				5.534

Communities affected:
No attempt has been made to estimate concentrations in populated areas.

Distances from releases to affected communities:
No attempt has been made to estimate concentrations in populated areas.

Weather conditions:
During BW 6-52, the wind was approximately south-southwest at 4 to 5 mph, carrying the FP over the sampling grid. During BW 5-52, the wind shifted and apparently carried the FP cloud west and then southwest out of the sampling grid.

Other materials released:
In BW 6-52, BG and SM were simultaneously released with the FP. In BW 5-52, AF and SM were simultaneously released with the FP.

Maximum time-integrated concentrations:
BW 6-52: 61 µg-min/m^3
BW 5-52: 16 µg-min/m^3

These were measured at 1 mile from the source. The wind shift during BW 5-52 probably resulted in a low measurement. In BW 6-52, the exposure was as high as 20 µg-min/m^3 at 13 miles from the source.

Maximum time-integrated concentrations in any populated area:
No attempt has been made to estimate concentrations in populated areas.

Maximum concentrations:
No time-resolved data were available to allow estimation of concentration.

Maximum concentrations in any populated area:
No attempt has been made to estimate concentrations in populated areas.

Other comments:
The north direction arrows on figures 6 and 7 are reversed with respect to the other figures, and with respect to reality. For various reasons, not discussed here, both these tests were complete failures from the point of view of their objectives.

APPENDIX B 167

DESCRIPTION OF TEST

Name of test: *Relationship of Dosages to Source Strength for BG and an Inert Tracer*

Reference-list number: 4

Reference: Trial Reports. Relationship of Dosages to Source Strength for BG and an Inert Tracer:
Report DPG BW 8A-1-54 24 Feb. 1954
Report DPG BW 8A-2-54 9 Mar. 1954
Report DPG BW 8A-4-54 22 Mar. 1954
Report DPG BW 8A-5-54 12 Apr. 1954

Principal object:
Not stated, except by implication in the title. The test plan (BW 8-54) was not available.

Site selection:
Dugway Horizontal grid.

Number and nomenclature of releases:
One release per report (four releases). The reports were numbered in the same way as the trials (BW 8A-1-54 through 8A-5-54, omitting 8A-3-54).

Test conditions:
Tests were according to a test plan that is not available (see "Principal object" above). A Stanford Blower was used, presumably for FP dissemination. Filter-type samplers were used.

Test material:
FP (nature unspecified). Particle counts were given as 1.96×10^{10} ppg for the last three experiments, and not specified for the first.

Place of release:
Dugway horizontal grid.

Dates, times, and quantities of release:

Test	Date	Start Time	Amount, g
BW 8A-1-54	1/21/54	15:54	9.69
BW 8A-2-54	1/27/54	15:59	7.935
BW 8A-4-54	2/17/54	19:15	8.966
BW 8A-5-54	3/14/54	19:02	8.23
Total			34.8

The dissemination time is not specified, but was probably short, since the experiments were apparently attempting to compare dispersion of FP with BG from bombs.

Communities affected:
No attempt has been made to estimate concentrations in populated areas.

Distances from releases to affected communities:
No attempt has been made to estimate concentrations in populated areas.

Weather conditions:

Test	Wind Speed at 2 m, mph	Wind Direction, deg.	Temp. at 2 m,°C	Rel. Hum., %
BW 8A-1-54	4	158	34.5	64
BW 8A-2-54	13	145	34.6	64
BW 8A-4-54	12.4	145	45.9	29
BW 8A-5-54	5.7	149	49.5	38

Maximum time-integrated concentrations:
No calibration data were provided, only the raw counts on Millipore filters. No information on the samplers was available. The following estimates of total exposure assume a sampling rate of 6.7 L/min, identical to that in the releases in May and June of 1953. The particles-per-gram count for the first experiment has been assumed to be identical to that for the other three experiments.

APPENDIX B

Test	Max. Exp., μg-min/m^3
BW 8A-1-54	94
BW 8A-2-54	270
BW 8A-4-54	250
BW 8A-5-5	1390

Maximum concentrations:
No time-resolved data, or even release-timing data, were available to allow estimation of concentration.

Maximum time-integrated concentrations in any populated area:
No attempt has been made to estimate concentrations in populated areas.

Maximum concentrations in any populated area:
No attempt has been made to estimate concentrations in populated areas.

Other comments:
All data are noted as preliminary and subject to revision. Each is a 9-10-page report of a release, including raw counts for organisms and FP.

The nomenclature suggests that a report for trial 8A-3-54 is missing.

DESCRIPTION OF TEST

Name of test: *Operation "Trouble Maker"*

Reference-list number: 6

Reference: Trial Report. BWAL, BW 8B-7 and 8-54, Operation "Trouble Maker." Relationship of dosages to source strength for BG and an inert tracer. BWALTR 27, 18 July 1955.

Principal object:
Not specified. The test plan (test plan BW 8B-54, 29 Nov. 1954, 6 pp., amended 12 Jan. 1955, 4 pp.) was not available.

Site selection:
Dugway west vertical grid.

Number and nomenclature of releases:
Two releases: BW 8B-7-54 and BW 8B-8-54.

Test conditions:
Tests were conducted according to a test plan that is not available (see "Principal object," above). A total of 318 Millipore filters were used for FP collection. A Skil blower was used, presumably for FP dissemination.

Test material:
FP (not otherwise specified). Particle count estimated as 2×10^{10} ppg.

Place of release:
Dugway west vertical grid.

Dates, times, and quantities of release:

Test	Date	Start Time	Amount, g
BW 8B-7-54	5/18/55	17:35	21.2
BW 8B-8-54	5/18/55	19:24	21.2
Total			42.4

No information on dissemination times is provided. BG was dispersed (presumably simultaneously) from bomblets; therefore, the dissemination time is likely to have been short.

Communities affected:
No attempt has been made to estimate concentrations in populated areas.

Distances from releases to affected communities:
No attempt has been made to estimate concentrations in populated areas.

Weather conditions:
Wind was approximately from the north, with a speed of about 16 mph

APPENDIX B

(at 2 m) during BW 8B-7-54. It was northerly with a speed of about 11 mph (at 2 m) during BW 8B-8-54.

Other materials released:
BG was released from bomblets.

Maximum time-integrated concentrations:
No calibration data were provided, only the raw counts on Millipore filters. No information on the samplers was available. The following estimates of total exposure assume a sampling rate of 6.7 L/min, identical to that in the releases in May and June of 1953.

BW 8B-7-54: Maximum exposure, 180 µg-min/m^3
BW 8B-8-54: Maximum exposure, 420 µg-min/m^3

Maximum concentrations:
No time-resolved data, or even release-timing data, were available to allow estimation of concentration.

Maximum time-integrated concentrations in any populated area:
No attempt has been made to estimate concentrations in populated areas.

Maximum concentrations in any populated area:
No attempt has been made to estimate concentrations in populated areas.

Other comments:
This report was a 20-page summary, subject to revision, giving raw counts and summary meteorologic data only.

DESCRIPTION OF TEST

Name of test: *Operation DEW I*

Reference-list number: 8

Reference: An Experimental Study of Long Range Aerosol Cloud Travel. Special Report 162. U.S. Army Chemical Corps Biological Laboratories, Camp Detrick, Frederick, Md. 1 Aug. 1952.

Principal object:
To test the possibility of achieving long-range aerosol-cloud travel and consequent coverage of large areas (several thousand square miles) at ground level beneath a slow-moving frontal system.

Site selection:
The southeastern states of North Carolina, South Carolina, and Georgia were selected as a test area after "consideration of meteorologic situations, terrain and modes of dissemination."

Number and nomenclature of releases:
Five releases, named Trial 1 through Trial 5.

Test conditions:
Releases were from a naval surface craft (USS TERCEL, a converted mine sweeper) operating at 15 knots parallel to the coast and about 5 to 10 miles offshore on a line segment approximately 100-150 nautical miles long off the North Carolina, South Carolina, or Georgia coasts on the total 390 nautical miles track between Jacksonville, Fla. and Hatteras, N.C. Stanford-type aerosol generators were used. A total of 46 sampling stations were established in the coastal plain area of Georgia, South Carolina, and North Carolina; sampling was set up for 15 consecutive 2-h periods during each trial.

Trial 1 occurred under conditions that were not correctly anticipated, so that the plume traveled principally parallel to the coast and out to sea. Trials 2 through 4 achieved the objectives, with the plumes traveling over large areas of Georgia, South Carolina, and North Carolina.

APPENDIX B

Test material:
ZnCdS (New Jersey Zinc Company, No. 2266), particle size 2.25 μm mass median diameter. Mean particle count approximately 4.2×10^{10} ppg (see "Other comments" below). Examination of this material also showed about a 1% contamination with soluble particles that fluoresced very intensely green, but were not otherwise identified.

Place of release:
For all trials, along a track about 5-10 miles offshore, between Jacksonville, Fla. and Hatteras, N.C.

Trial 1: Along 138 statute miles, starting approximately 35 statute miles from Jacksonville, northward direction of travel.
Trial 2: Along 121 statute miles, starting approximately 121 statute miles from Jacksonville, northward direction of travel.
Trial 3: Along 121 statute miles, starting approximately 173 statute miles from Jacksonville, southward direction of travel.
Trial 4: Along 121 statute miles, starting approximately 331 statute miles from Jacksonville, southward direction of travel.
Trial 5: Along 156 statute miles, starting approximately 173 statute miles from Jacksonville, southward direction of travel.

Dates, start times, and quantities of release:
Trial 1: 3/26/52   0600-1400 EST 240 lb (109 kg, 4.6×10^{15} particles)
Trial 2: 3/30-31/52 1900-0200 EST 250 lb (113 kg, 4.8×10^{15} particles)
Trial 3: 4/4/52   1100-1800 EST 250 lb (113 kg, 4.8×10^{15} particles)
Trial 4: 4/9/52   1100-1800 EST 200 lb (91 kg, 3.8×10^{15} particles)
Trial 5: 4/21/52   1400-2300 EST 450 lb (204 kg, 8.6×10^{15} particles)
Total release to the environment: 1,390 lb (630 kg).

Communities affected:

The following describes areas affected at a cumulative dosage greater than 1 particle-min/L (approximately 2.7×10^{-2} µg-min/m^3). The areas are so large that only a general description can be given.

Trial 1: Approximately 1,000 square miles. Almost everywhere within approximately 5-10 miles of the coast of Georgia, and within a few miles of the southernmost 30 miles of the coast of South Carolina.

Trial 2: Approximately 21,800 square miles, covering approximately the southern half of South Carolina and an abutting swathe of east Georgia approximately 25 miles wide at the coast, widening to 75 miles at 150 miles inland.

Trial 3: Approximately 13,100 square miles, roughly encompassed by a north-south stripe across the central one-third of South Carolina.

Trial 4: Approximately 28,900 square miles covering the southern two-thirds of South Carolina, together with an abutting 60-mile-radius semicircle extending into Georgia, with its base on the state line and extending inland 120 miles from the coast.

Trial 5: Approximately 34,800 square miles, a coastal strip 75 miles wide from the southern most tip of the Georgia coast to Cape Hatteras, N.C.

Distances from releases to affected communities:

The releases were 5-10 miles offshore and 5-10 miles from the nearest affected communities. Measurable plume concentrations extended up to 175 miles inland.

Weather conditions:

The releases were designed to be released with onshore winds under frontal inversions and, taking advantage of nocturnal inversions, to ensure long-distance travel with minimum vertical dispersion. During Trial 1, the winds remained parallel to the shoreline instead of turning onshore as predicted, and so the plume missed most land. Trial 2 had onshore winds and nocturnal inversions throughout the area, but no frontal inversion. Low clouds formed during the early morning, followed by showers, light rain, and drizzle. The plume (disseminated from 1900 through 0200 h) was observed to increase in area until morning when the nocturnal inversions dissipated. Trial 3 achieved substantial dissemination in the cold air under a warm front that was moving northward, accompanied by unstable air with small tornadoes, followed by rain. Measurements indicated substantial trapping under the frontal system. In Trial 4, dissemination (1100-1800 h) was into a neutral atmosphere with a weak warm front that dissipated rapidly. Coastal clouds dissipated later in the day, with clear inland weather and onshore winds. A wind shift across the weak front apparently resulted in two effective plumes from different parts of the dissemination line. Trial 5 was conducted when no frontal conditions were present, but a radiation inversion formed and was present at night (dissemination 1400-2300 h). The early part of the release moved rapidly, but the later (nocturnal) part of the release was into relatively light winds in neutral conditions. The nocturnal plume spread slowly until morning, when it dissipated rapidly as the inversion was destroyed by daytime heating.

Other materials released:
There is no mention of other releases in the unclassified document.

Maximum time-integrated concentrations:
These were measured over 30 h from the beginning of dissemination, ensuring complete plume capture (except for the very minor re-entrainment from deposited material). Measurement accuracy was of the order of $\pm 40\%$, so the 2-digit precision given here is for calculation purposes only. The values shown might be underestimated by a factor of 7 (see "Other comments" below).

Trial 1: 0.33 μg-min/m^3
Trial 2: 16 μg-min/m^3
Trial 3: 1.0 μg-min/m^3
Trial 4: 5.8 μg-min/m^3
Trial 5: 91 μg-min/m^3

Combined: 98 µg-min/m³ (The maximum time-integrated concentrations for all trials at each individual measurement point.)

Maximum time-integrated concentrations in any populated area:
Same as the maximum time-integrated concentrations anywhere, because the dispersion area was so large. The highest cumulative concentration was measured at Hunter Air Force Base, Ga.

Maximum concentrations:
These concentrations are 2-h averages, as measured in the tests. Once again, accuracy is approximately ±40%, so the two digit precision given here is for calculation purposes only. The values shown might be underestimated by a factor of 7 (see "Other comments" below).

Trial 1: 1.3×10^{-3} µg/m³
Trial 2: 7.0×10^{-2} µg/m³
Trial 3: 8.1×10^{-3} µg/m³
Trial 4: 2.4×10^{-2} µg/m³
Trial 5: 3.4×10^{-1} µg/m³
Combined: 3.4×10^{-1} µg/m³ (The largest value found in any trial.)

Maximum concentrations in any populated area:
Same as the maximum concentrations anywhere, since the dispersion area was so large. Maximum concentration was measured at Hunter Air Force Base, Ga.

Other comments:
Estimates of deposition indicate that only 5% to 6% of the material was deposited within the measurable plume area (out to 150 miles).

The calculation of number of particles released was not discussed. It appears that the cited particle "mass median" diameter of 2.25 µm was used, together with a particle density of 4 g/cm³, and an assumption of spherical particles. This gives the approximate counts cited above, but

APPENDIX B 177

it ignores the efficiency of dissemination and the difference between mean and median diameters. (Note that "mass median" is a misnomer—this is also the number, surface area, and every other median diameter as well.) This could have led to an underestimate of a factor of approximately 7 in the concentration and exposure estimates given above (assuming number counts and dissemination efficiencies close to Ref. 19).

DESCRIPTION OF TEST

Name of test: *Project WINDSOC*

Reference-list number: 13

Reference: T.B. Smith and M.A. Wolf. Intermediate Scale Particulate Cloud Travel—Project WINDSOC—Part 1—Technical Report. Meteorology Research, 2420 N. Lake Ave., Altadena, Ca. MRI Report 60-23. Prepared for U.S. Army Chemical Corps Proving Ground under Contract DA-42-007-403-CMI-432. July 1960. (Only the first 18 pages of this over 73-page report were available.)

Principal object:

Informally:
"To study particulate cloud travel under night release conditions on a distance scale intermediate between the LAC operation and the small scale diffusion studies usually performed."

Formally:
Principal objective: "To describe and determine empirically the relationships between meteorologic forces acting on elevated line sources of particulate aerosols and the resulting quantitative distribution of those particulates at ground level."

Secondary objective: "To provide data for the extension of existing dosage predictive techniques, or the development of new techniques, appropriate for the aerial line-source dissemination of particulates."

Number and nomenclature of releases:
A total of 13 tests using FP tracer. They are referred to as Test 1 through Test 13.

Test conditions:
Release was into stable air over relatively flat terrain in central Texas. The test area (125 × 125 miles south and west of Dallas, including Dallas, Fort Worth, Waco, and Fort Hood) was selected for such weather and terrain characteristics. A total of 84 ground samplers were used in the test area, together with balloon samples at Fort Hood for some tests, five samplers (at 300, 600, 900, 1,200, and 1,400 ft above ground) on a TV tower 20 miles southwest of Dallas for Tests 9 through 13, groups of Rotorod samplers at various points on the ground for various tests, and samplers aboard four L-20 aircraft. FP tracer was released from a C-119 aircraft flying crosswind along the upwind side of the test area, at a height of 600 to 1,300 ft. above ground level, along 200 miles (i.e., extending well beyond the edges of the test area. Tests 5, 9 and 11 were not fully analyzed, since meteorologic conditions did not meet requirements.

Test material:
ZnCdS from U.S. Radium Corp. (Contract DA-18-064-404-CML-427). Average particle count per gram of five lots was 2.09×10^{10}; differences between lots were not considered meaningful. The mode of the particle diameter was near 2 μm, with 10-25% outside the range 1-4 μm.

Place of release:
Along a 200-mile line over-running the 125-mile upwind edge of the test area.

Dates, times, and quantities of release:

Appendix B

The release rate varied during each test, with a maximum of 4.5 lb/mile (Test 1). From Test 3 onward, it was stabilized at approximately 3.2 lb/mile. The implied total releases were thus:

Test 1:	Aug. 1959	900 lb
Test 2:	Aug. 1959	640 to 900 lb
Test 3:	Aug. 1959	640 lb
Test 4-8	Oct. 1959	640 lb
Test 9-13	Feb. 1960	640 lb

Other information is unavailable in the section of report obtained.

Communities affected:
All those within the 125 × 125 mile test area, including Dallas, Fort Worth, Fort Hood, and Waco. In addition, communities outside the test area but within the plume from the 200-mile release.

Other information is available in the section of the report obtained.

Distances from releases to affected communities:
Unavailable in the section of the report obtained.

Weather conditions:
Unavailable in the section of the report obtained.

Other materials released:
Unavailable in the section of the report obtained.

Maximum time-integrated concentrations:
Unavailable in the section of the report obtained.

Maximum time-integrated concentrations in any populated area:
Unavailable in the section of the report obtained.

Maximum concentrations:
Unavailable in the section of the report obtained.

Maximum concentrations in any populated area:
Unavailable in the section of the report obtained.

Other comments:
The full report contains all the information unavailable in the 18 pages that are available. Mention is made in this report of a series of 11 trials at Dugway, 21-23 Sept. 1959, to test the dissemination efficiency of the plane. Measured particle number efficiency was 13% to 72%.

DESCRIPTION OF TEST

Name of test: *Vertical Diffusion from an Elevated Line Source* (no code name identified)

Reference-list number: 16

Reference: T.B. Smith and M.A. Wolf. Vertical Diffusion from an Elevated Line Source over a Variety of Terrains. Part A. Final Report to Dugway Proving Ground. Meteorology Research, 2420 North Lake Ave., Altadena, Calif. Contract DA-42-007-CML-545. 31 Mar. 1963

Principal object:
Determination of the applicability of the dosage prediction technique of the Dallas Tower study under differing conditions of other areas.

Site selection:
All four selected sites were required to be in areas where adverse conditions (rain, fog, low clouds, extreme wind speeds, variable wind direction) were minimized, and where there were existing roads parallel to the persistent wind direction. Two sites, in Oklahoma and Washington, were selected on the basis of their homogeneous rolling terrain (providing greater surface roughness than the Dallas Tower study). The two other sites were selected for single features—the Texas site offers a flat coast

line normal to the wind flow; the Nevada site presents a cross-wind ridge on an otherwise smooth surface.

Number and nomenclature of releases:
A total of 36 releases, identified as Test 1 through Test 36, with three considered unsuccessful (Test 10, disseminator malfunction; Tests 20 and 34, wind shifts). There were nine releases at the first three sites (Oklahoma, Tests 1 through 9; Texas, Tests 11 through 19; Washington, Tests 20 through 28), and eight at the last site (Nevada, Tests 29 through 36).

Test conditions:
At each site, the releases were made from an Aero Commander airplane flying at 500 to 1300 ft. FP were metered for release at 1.5 lb/mile over a flight path of 30 miles, yielding a total mass released of 45 lb per release. Sampling was performed at 1-mile intervals along approximately 25 miles in the downwind direction, generally starting a short distance upwind of anticipated initial touchdown point of the plume.

Test material:
ZnCdS containing 1% by weight of micronized Valron Estersil for increased fluidity (Lot 14, produced Nov. 1959 by U.S. Radium Corp.). Median diameter approximately 2.4 μm. Mean particle count 2.16×10^{10} per gram, with a release efficiency assumed to be the same (39%) as in the Dallas tests, yielding a particle emission rate of 1.11×10^9 per foot.

Place of release:
Oklahoma: Along 30-mile approximately east-west tracks, 4 to 5 miles south of the line joining Ripley, Cushing, and Drumright, centered on Cushing (and centered on Route 18, along which, to the north of Cushing, the sampling points extended).

Texas: Along 30-mile approximately northeast-southwest tracks, 2 to 4 miles out into the Gulf of Mexico off Mustang Island, centered on Port Aransas. The sampling points were located along the causeway from Port Aransas to Aransas Pass, continuing along the road to Gregory and then to Taft.

Washington: Along 30-mile approximately east-west tracks, within approximately 2 miles north or south of Colfax, centered on Colfax. There were two locations for the line of samplers, both initially heading north-northeast from Colfax along the main road, the first then turning north to Rosalia, the second continuing north-northeast through Oakesdale to Tekoa.

Nevada: Along 30-mile approximately east-west tracks, approximately 11 to 12 miles north of Goldfield, and centered on Route 95. The line of samplers extended along Route 95 from approximately 11 miles north of Goldfield to approximately 11 miles south of Goldfield.

Dates, times, and quantities of release:

Test	Date	Local time	Quantity
Oklahoma			
1	06/04/62	18:13	50 lb (22.7 kg)
2	06/04/62	22:16	50 lb (22.7 kg)
3	06/05/62	20:42	50 lb (22.7 kg)
4	06/14/62	20:57	50 lb (22.7 kg)
5	06/15/62	19:52	50 lb (22.7 kg)
6	06/15/62	23:42	50 lb (22.7 kg)
7	06/16/62	03:48	50 lb (22.7 kg)
8	06/16/62	20:05	50 lb (22.7 kg)
9	06/16/62	23:10	50 lb (22.7 kg)
Total			450 lb (204 kg)
Texas			
11	06/24/62	16:12	50 lb (22.7 kg)
12	06/24/62	20:09	50 lb (22.7 kg)
13	06/25/62	00:01	50 lb (22.7 kg)
14	06/27/62	19:44	50 lb (22.7 kg)
15	06/28/62	19:38	50 lb (22.7 kg)

APPENDIX B 183

16	06/28/62	23:38	50 lb (22.7 kg)	
17	06/29/62	19:28	50 lb (22.7 kg)	
18	06/29/62	23:37	50 lb (22.7 kg)	
19	06/29/62	03:28	50 lb (22.7 kg)	
Total			450 lb (204 kg)	

Test	Date	Local time	Quantity
Washington			
20	10/02/62	22:15	50 lb (22.7 kg)
21	10/06/62	14:52	50 lb (22.7 kg)
22	10/06/62	18:07	50 lb (22.7 kg)
23	10/06/62	22:19	50 lb (22.7 kg)
24	10/07/62	01:36	50 lb (22.7 kg)
25	10/15/62	18:16	50 lb (22.7 kg)
26	10/21/62	14:53	50 lb (22.7 kg)
27	10/21/62	18:05	50 lb (22.7 kg)
28	10/21/62	22:18	50 lb (22.7 kg)
Total			450 lb (204 kg)
Nevada			
29	10/31/62	15:55	50 lb (22.7 kg)
30	10/31/62	19:18	50 lb (22.7 kg)
31	11/01/62	19:08	50 lb (22.7 kg)
32	11/01/62	22:02	50 lb (22.7 kg)
33	11/02/62	02:02	50 lb (22.7 kg)
34	11/04/62	17:33	50 lb (22.7 kg)
35	11/05/62	19:03	50 lb (22.7 kg)
36	11/05/62	21:54	50 lb (22.7 kg)
Total			400 lb (181 kg)

Communities affected:

The nearest communities (on current maps) that were potentially most affected were the following:

Oklahoma: Cushing, Ripley, Drumright, Oilton, Yale, Jennings, Hallett, Glencoe, Pawnee, and nearby towns, principally those to the north.

Texas: Port Aransas, Aransas Pass, Ingleside, Gregory, Portland, Corpus Christi, Rockport, Bayside, Taft and nearby towns, principally those to the northwest.

Washington: Colfax, Palouse, Garfield, Steptoe, Thornton, Oakesdale, Malden, Rosalia, Tekoa, Farmington, Latah, and nearby towns, principally those to the north and north-northeast.

Nevada: Goldfield, and other towns to the south.

Distances from releases to affected communities:
Oklahoma: Releases were approximately 4 or 5 miles upwind of Ripley, Cushing, and Drumright. Other towns were at least 5 miles further downwind.

Texas: Releases were approximately 2 to 5 miles from Port Aransas and at least 8 miles from other communities.

Washington: Releases were mostly slightly north (downwind) of Colfax, with apparently one release slightly (a mile or so) upwind of Colfax. Most releases were probably about 5 miles upwind of Palouse and more than 8 miles upwind of other communities.

Nevada: Releases were approximately 4 to 11 miles north (upwind) of Goldfield. Other towns were at least 30 miles downwind of the release.

Weather conditions:
In all cases, the tests were carried out in the desired clear weather, with specific meteorologic regimes that established relatively high turbulence in the lower layers of the atmosphere.

Oklahoma: Tests were conducted in typical summer weather, with southerly flow conditions (wind direction recorded as south two times, south-southeast four times, and southeast three times), when turbulence is generated in the lower 1,000-1,500 ft, with little turbulence above 1,000-1,500 ft. In Test 4, however, there was insufficient vertical wind shear to generate turbulence in the layers near the release height.

Texas: Tests were conducted with south to southeasterly flow conditions (wind direction recorded as southeast five times, south-southeast two times, and east-southeast two times). Solar heating and ocean conditions were such that the turbulence in the lower 1,000-1,500 ft was relatively low compared with the other test areas.

Washington: Tests were conducted in south to southwesterly wind flow conditions (wind direction recorded as south-southeast three times, south one time, south-southwest three times and southwest two times). The desired relatively high turbulence in the lower levels of the atmosphere is not typical of the area, and a 3-week period was required to complete the tests.

Nevada: The tests were conducted in the steady north to northwest flow regime that often develops in the late afternoon and night (recorded wind directions were northeast one time, north five times, and north-northwest two times). This wind flow developed despite easterly pressure-gradient winds (above 7,000 ft), and has a depth of about 2,000 ft, passing easily over the Goldfield ridge.

Other materials released:
No other materials are mentioned in the report.

Maximum time-integrated concentrations and maximum concentrations:
The report gives the measured cumulative exposure. The concentration estimates here were made by assuming a gaussian plume shape with along-wind dispersion of 1,000 m (corresponding approximately to C stability at 10 to 20 km downwind), and using the average measured wind speed at 10 m height. These estimates should thus be considered very rough estimates. Accuracy of measurements was not discussed. By com-

parison with other reports, the measurement accuracy cannot be better than about a factor of 2 (so the precision of the entries in the table should be retained only for checking calculations).

Test	µg-min/m³	µg/m³	Test	µg-min/m³	µg/m³
Oklahoma			Washington		
1	7.19	0.57	20	0.08	0.03
2	7.98	0.70	21	2.15	0.30
3	4.77	0.39	22	0.99	0.06
4	8.67	0.50	23	0.09	0.003
5	9.33	0.75	24	1.04	0.06
6	7.37	0.58	25	0.47	0.01
7	7.09	0.63	26	4.57	0.21
8	6.40	0.71	27	0.01	0.0003
9	4.37	0.53	28	0.01	0.001
All	39.3	0.75	All	6.72	0.30
Texas			Nevada		
11	3.41	0.74	29	21.4	1.6
12	5.08	0.39	30	4.34	0.29
13	3.18	0.32	31	1.27	0.14
14	*	*	32	5.76	0.37
15	*	*	33	2.65	0.11
16	5.59	0.47	34	0.12	0.003
17	5.73	0.54	35	2.88	0.46
18	6.81	0.61	36	1.90	0.16
19	9.79	0.82			
All	35.5	0.82	All	23.2	1.6

* = page missing in report. Total (All) integrated exposure estimates have been estimated by assuming that these two tests were equivalent to the averages of the other tests.

Maximum time-integrated concentrations and maximum concentrations in any populated area:

In Oklahoma, Texas, and Washington, the density of communities is high enough in the affected area that it is likely that populated areas experienced concentrations and time-integrated concentrations as high as were

APPENDIX B

measured anywhere. In Nevada, the only nearby community was the town of Goldfield, which experienced time-integrated concentrations (and probably maximum concentrations) between one-third and one-fourth of the maxima reported above for Nevada.

DESCRIPTION OF TEST

Name of test: *Tracer Study of Dispersion over a City* (no code name identified).

Reference-list numbers: 17 and 43

Reference: F. Pooler. A Tracer Study of Dispersion over a City. Paper 6628 at the 59th Annual Meeting, Air Pollution Control Association, San Francisco, Calif.; June 20-24, 1966 (from the Air Resources Field Research Office, ESSA, at the Robert A. Taft Sanitary Engineering Center, Division of Air Pollution, U.S. Department of Health, Education, and Welfare, Cincinnati, Ohio). Also published in 1966 in the *Journal of the Air Pollution Control Association* (Vol. 16, pp. 627-631). (The journal article appears to have been slightly edited.)

Principal object:
To obtain "direct experimental evidence to indicate at least order-of-magnitude urban area dispersion parameters."

Site selection:
"St. Louis was chosen as the experimental area for study primarily because a more general study of air pollution in the St. Louis metropolitan area was then being initiated. In addition," the city is reasonably flat and not affected by significant topographic features, the Weather Bureau Office operated a weather radar, and St. Louis was easily accessible from Cincinnati (experimenter's home office).

The areas principally affected by releases discussed in this reference are distinct from the areas affected by releases discussed in Ref. 35.

Number and nomenclature of releases:
Seven series of experiments, with a total of 42 releases. Each release is labeled by a series number (1 through 7) and a test number from 2 to 43. (Test 1 was a dry run with no tracer release.)

Test conditions:
Samplers used included 60 Rotorods, 30 membrane filters, and 10 pulsed drum samplers. A total of 316 sampling sites were used during some or all tests, placed in arcs around the sources at nominal distances of 0.5, 2, and 4.5 miles from Forest Park, and 1.25, 2.5, and 5 miles from the Knights of Columbus Building (see below). Two generators were available that were capable of release rates from 0.5 to 200 g/min. Forty of the 42 tests had a 1-h release time, so that the maximum release possible was 24 kg per test (2×0.2 kg/min $\times 60$ min); the other two tests had a release period of 0.5 h, giving a maximum possible release of 12 kg.

Test material:
Not discussed, except by reference to the literature.

Place of release:
Releases were from a site in Forest Park (Tests 2-4, 9, 16-20, 22-25, 29, 31-41, 43) or from the roof of the Knights of Columbus Building. The available map is of low resolution, but places the release sites in the vicinity of the junction of Clayton Road and Faulkner Road in Forest Park and near the junction of S. Grand Blvd. and Gravois Ave. for the Knights of Columbus Building. Those two sites are approximately 2.8 miles apart.

Dates, times, and quantities of release:
Dates, day of week, times, and source location were the following:

APPENDIX B

Series	Test	Date	Day	Time, CST Begin	End	Source Location[a]
1	2	05/27/63	Mon	14:10	14:40	A
	3	05/28/63	Tue	10:00	11:00	A
2	4	07/19/63	Fri	11:30	12:30	A
	5	07/22/63	Mon	11:00	12:00	B
	6	07/23/63	Tue	11:30	12:30	B
	7	07/25/63	Thu	10:40	11:40	B
	8	07/26/63	Fri	10:45	11:45	B
3	9	09/12/63	Thu	11:15	12:15	A
	10	09/14/63	Sat	10:45	11:45	B
	11	09/16/63	Mon	11:00	12:00	B
	12	09/17/63	Tue	20:00	20:30	B
	13	09/18/63	Wed	20:00	21:00	B
4	14	04/01/64	Wed	12:00	13:00	B
	15	04/06/64	Mon	20:40	21:40	B
	16	04/07/64	Tue	20:48	21:48	A
	17	04/08/64	Wed	20:30	21:30	A
	18	04/09/64	Thu	20:45	21:45	A
5	19	06/02/64	Tue	10:30	11:30	A
	20	06/03/64	Wed	10:40	11:40	A
	21	06/04/64	Thu	10:30	11:30	B
	22	06/06/64	Sat	11:30	12:30	A
	23	06/07/64	Sun	11:30	12:30	A
	24	06/09/64	Tue	10:30	11:30	A
	25	06/10/64	Wed	10:30	11:30	A
	26	06/11/64	Thu	10:30	11:30	B
6	27	10/10/64	Sat	11:30	12:30	B
	28	10/11/64	Sun	11:05	12:05	B
	29	10/12/64	Mon	20:00	21:00	A
	30	10/16/64	Fri	20:00	21:00	B
6	31	10/17/64	Sat	13:15	14:15	A
	32	10/19/64	Mon	19:45	20:45	A
	33	10/20/64	Tue	19:15	20:15	A
	34	10/21/64	Wed	19:00	20:00	A
7	35	03/06/65	Sat	12:30	13:30	A
	36	03/07/65	Sun	12:30	13:30	A
	37	03/08/65	Mon	20:30	21:30	A
	38	03/11/65	Thu	20:30	21:30	A

(Continued)

Series	Test	Date	Day	Time, CST Begin	End	Source Location[a]
	39	03/13/65	Sat	12:20	13:20	A
	40	03/14/65	Sun	11:00	12:00	A
	41	03/15/65	Mon	20:50	21:50	A
	42	03/16/65	Tue	20:30	21:30	B
	43	03/17/65	Wed	20:00	21:00	A

[a]Source locations: A = Forest Park; B = Knights of Columbus Building.

Quantities were not given. From the data presented in the paper for Experiment 9, it is possible to infer a release quantity of approximately 24 kg for that experiment, corresponding to the maximum dispersal rate of the two generators available and assuming 1×10^{10} ppg for the release rate. The concentration data presented for Experiments 18 and 32 are consistent with a similar release quantity.

Experiment 9 is presented as typical of daytime releases (26 cases) and Experiments 18 and 32 of evening experiments (16 cases). If all experiments were carried out at the maximum dispersal rate, the total release would have been approximately 984 kg (40 experiments had a 1-h dispersal period, and two had a 0.5-h dispersal period).

Communities affected:
St. Louis.

Distances from releases to affected communities:
Release took place within St. Louis.

Weather conditions:
The intent was to run tests during various weather conditions. There is no discussion of actual weather conditions, although windy conditions were noted in one table, and there is a notation that light snow ended during Experiment 36.

APPENDIX B   191

Other materials released:
No other material is mentioned.

Maximum time-integrated concentrations:
Very few data were provided in this short paper.

Nominal particle concentrations, obtained by dividing measured exposures by the release times, were given graphically for three experiments (numbers 9, 18, and 32). The maximum values were approximately 2×10^5 particles per cubic meter, 4×10^5 particles per cubic meter, and 1×10^5 particles per cubic meter, respectively. If the material used was similar to that used in other experiments (about 1×10^{10} particles released per gram), those values correspond to average concentrations of 20, 40, and 10 µg/m³, respectively, or exposures of 1,200, 2,400, and 600 µg-min/m³, respectively, for 1-h releases. The affected area in each case covered approximately a 40-degree arc of receptors, and the receptor locations allowed coverage of approximately three quadrants when both sources are taken into account.

Experiment 9 was said to be typical of daytime conditions (26 experiments, 15 from Forest Park) and Experiments 18 and 32 of evening conditions (16 experiments, 11 from Forest Park). Taking account of different wind directions (covering the approximately 180 degrees of good coverage by the nearest receptor arcs) during the various releases from Forest Park, the highest cumulative exposure in the vicinity of Forest Park was almost certainly less than 7,400 µg-min/m³.

Maximum time-integrated concentrations in any populated area:
Same as the maximum time-integrated concentrations anywhere.

Maximum concentrations:
Not computable from information currently available.

Graphs in the paper indicate particle concentrations for three tests, the highest value of about 4×10^5 particles per cubic meter being in Test 18 at the closest measurement points, approximately 640 m from the release point. If the material used was similar to that used in other experiments (about 1×10^{10} particles released per gram), that value corresponds to a concentration of 40 μg/m³.

Maximum concentrations in any populated area:
Same as the maximum concentrations anywhere.

Other comments:
There should be much more information available about this test in files of the U.S. Department of Health and Human Services. The paper has some graphs of "concentrations" in particles per cubic meter (see "Maximum time-integrated concentrations" and "Maximum concentrations" above), although the basis for these graphs is not given. Other bar charts show particle counts per minute, presumably obtained from the drum-pulsed samplers mentioned. However, no raw data were provided in the paper, and much of the information needed for accurate estimates is missing.

DESCRIPTION OF TEST

Name of test: *Calibration of the Model D-1 Dry Particulate Disseminator, DPGTP 508C*

Reference-list number: 18

Reference: Calibration of the Model D-1 Dry Particulate Disseminator with Green Fluorescing FP, DPGTP 508C. USATECOM Project 5-3-9030-10. DPGTM 1060. Biological Branch, Test Design and Analysis Division, Technical Plans and Evaluation Directorate, U.S. Army Test and Evaluation Command, Dugway Proving Ground, Dugway, Utah. Case File 3164. Nov. 1963.

Principal object:
To determine estimates of dissemination efficiency and source strengths

APPENDIX B   193

achieved with green fluorescing FP when disseminated from the Model D-1 (L-20) dry particulate disseminator.

Site selection:
Aerial Spray Grid, located 11 miles west of the technical area, Dugway Proving Ground.

Number and nomenclature of releases:
The whole series of nine trials constituted DPGTP 508C. There was one release per trial, numbered C-1 through C-8, with an interpolated C-2R. C-2 was aborted because of malfunctioning sampling equipment and was repeated as C-2R.

Test conditions:
A U-6 aircraft disseminated the FP while flying at 130 to 175 ft altitude at 115 mph along a 15,000-ft flight line 100 yd upwind from the 300-ft tower of the aerial spray grid (ASG). Dissemination was at 7.5 lb/min for approximately 1.5 min. The ASG was equipped with 58 Millipore filters spaced 5 ft vertically apart up the tower.

Test material:
Green fluorescing ZnCdS FP, Lot H-324, from U.S. Radium Corp., with 1% wt/wt micronized DuPont Valron Estersil, particle count 3.59×10^{10}, according to an inspection report by Laboratories Branch, Biological Division.

Place of release:
See "Site selection" and "Test conditions" above.

Dates, times, and quantities of release:

Test	Date	Time	Duration, s	Amount, g	At 150 ft Wind Speed mph	Direction Deg.
C-1	05/17/63	12:52	87	3607	5	360
C-2	No further data available					
C-2R	05/21/63	13:30	70	2894	7	290
C-3	No further data available					
C-4	05/20/63	17:26	82	3394	10	356
C-5	05/20/63	18:21	81	3164	7.8	350
C-6	05/21/63	14:28	72	2920	6.5	270
C-7	08/15/63	11:27	78	3512	5.3	293
C-8	08/15/63	13:12	82	3532	4.7	356
	Adjusted total			29,601		

Trial C-2 was aborted, and no data were presented. In trial C-3, the wind angle changed at the last moment, and the data were not analyzed or presented. The total amount released has been increased to account for the missing data, by assuming dissemination of the average of the seven available amounts.

Communities affected:
No attempt has been made to estimate concentrations in populated areas.

Distances from releases to affected communities:
No attempt has been made to estimate concentrations in populated areas.

Weather conditions:
See "Dates, times, and quantities" above for wind angle and speed half way up the Aerial Spray Grid tower (at 150 ft).

Other materials released:
No other materials were mentioned.

Maximum time-integrated concentrations:
The maximum concentrations listed below take account of the efficien-

cies of dissemination measured in these tests (average about 50%). These maximum concentrations were measured at heights varying from ground level to 175 ft, with measurements 100 yd downwind of the releases. The cumulative maximum measured was at 175 ft, although nearly the same cumulative total was recorded at ground level.

Test	Max. Exp., μg-min/m^3
C-1	756
C-2	No data available
C-2R	316
C-3	No data available
C-4	191
C-5	342
C-6	603
C-7	260
C-8	445
Cumulative	1044

Maximum time-integrated concentrations in any populated area:
No attempt has been made to estimate concentrations in populated areas.

Maximum concentrations:
No time-resolved data were available to allow estimation of concentration.

Maximum concentrations in any populated area:
No attempt has been made to estimate concentrations in populated areas.

Other comments:
This was the third phase of the 508 series of efficiency estimates for FP dissemination. Test Series 508, Phases A and B involved at least five and three releases, respectively. In addition, the Bio 595 (SD-2 drone) program involved at least five releases. All the Bio 595 releases were of yellow fluorescing FP. The other releases are described in the reports:

DPGTM 1045, Field Calibration of the L 23 FP Disseminator, Bio 508A, Dugway, Utah. Jan. 1961.

DPGTM 1052, Calibration of the Model D-1 Dry Particulate Disseminator Designed for the L-20 Aircraft, Bio 508B, Dugway, Utah. July 1962.

DPGTM 1058, Drone Delivery System AN/USD-2 (XAE-3), Phase III, Biological Studies, Dugway, Utah. Aug. 1963.

DESCRIPTION OF TEST

Name of test: *Deciduous Forest Diffusion Study*

Reference-list number: 19

Reference: Deciduous Forest Diffusion Study. Report 3004. Final Report. Applied Science Division, Litton Systems, Minneapolis, Minnesota. Contract DA-42-007-AMC-48(R), Project 1T062111A128, for the Meteorology Division, Deseret Test Center, Building 103 Soldiers Circle, Fort Dugway, Utah. June 1969.
    Vol. 1: M.H. Tourin, W.C. Shen. Diffusion Studies in a Deciduous Forest.
    Vol. 2: M.H. Tourin, R.J. Rickett, W.C. Shen. Detailed Program Description.
    Vol. 3: M.H. Tourin, W.C. Shen. Detailed Technical Data Supplement.

Principal object:
To select an appropriate site, design a test grid, acquire pertinent meteorologic and aerosol-sampling data, and analyze these data in terms of meaningful diffusion characteristics related to a deciduous forest area and its associate micrometeorology.

Site selection:
Site selection for level, nonmarshy terrain with uniform, dense coverage of hardwood or deciduous trees; continuous site at least for 3 miles up-

wind; a minimum of hills, streams, lakes, built-up areas for several miles; presence (but not excessive) of snow in winter; minimum objections from private landowners and populated communities or interference from campers and public; accessibility by road; access to manpower, supplies, and facilities; reasonable convenience of aircraft facilities; and access to meteorologic data for previous years. Final site selected from possible test sites in Minnesota principally by Army helicopter survey. Site is within Chippewa National Forest in North Central Minnesota, approximately 350 km north of Minneapolis-St. Paul, 5-6 km from main roads, 8-10 km southeast of nearest town of Cass Lake (population 1,500), 32 km southeast of Bemidji (population 10,000).

Number and nomenclature of releases:
Two series of 12 tests were labeled Tests 102 through 113 (winter series) and 201 through 212 (summer series).

Test conditions:
A test array of Rotorod and other samplers was set up on towers and poles covering 2 km by 500 m, aligned northwest-southeast, with an especially dense microgrid at the center of the test array. FPs were released from a Piper Apache aircraft flying at 150 mph ground speed over a line length of 16 km (4-min flight time) approximately 1.5 km or 4.8 km (Tests 110 and 202) upwind of the closest tower of, perpendicular to and symmetric about the center-line of the test array, at the lowest feasible height (38 to 60 m, depending on conditions; tree height was 16-18 m); in addition, individual bomblets (all summer tests, and two winter tests) were exploded in the center of the central microgrid to simulate point sources, and a line of 9 bomblets (some winter tests) spaced at 5-m intervals through the microgrid, perpendicular to the axis of the test array, were used to simulate a line source. Samples were taken with 15-min integration time at heights of 1.5 (all test array points), 5, 10, 20, and 40 m.

Release rate from the aircraft was metered to 0.85 kg/km, so that total release per test was approximately 13.6 kg. Bomblets contained 25 (± 0.1) g.

Test material:
ZnCdS provided by Dugway Proving Ground.

Yellow fluorescing (aerial release): Type 2267, Lot H390, 1.56×10^{10} ppg; and Lot H395, 1.32×10^{10} ppg. Release efficiency taken to be 39%.

Green fluorescing (ground level bomblets): Type 3206, Lot H391, 1.62×10^{10} ppg. Loaded in modified E-61R4 bomblets that produce an instantaneous 2-m diameter spherical cloud on detonation. Release efficiency taken to be 30%.

Place of release:
For most tests, the aerial release was along a line 16 km long, perpendicular to and symmetric about, the main northwest-southeast axis of the test array, and 1.6 km northwest of its nearest point. For tests 103, 206, 208, the wind was from the opposite direction, and the release line was 1.6 km southeast of the nearest point of the test array. For tests 110 and 202, the release line was 4.8 km northwest of the test array. That is summarized in the table below.

Dates, times, and quantities of release:

Test	Date	Time, LST	Amount Released Air, kg	Ground, kg	Direction/ Distance, km
102	01/25/64	21:33	13.1	0.025	NW 1.6
103	01/30/64	01:06	13.8	0.025	SE 1.6
104	02/06/64	21:12	14.0	0.225	NW 1.6
105	02/12/64	21:11	13.9	0.225	NW 1.6
106	02/14/64	21:08	13.4	0.225	NW 1.6
107	02/18/64	20:44	13.5	0.225	NW 1.6
108	03/02/64	21:02	14.3	0.225	NW 1.6
109	03/06/64	20:55	13.4	0.225	NW 1.6
110	03/10/64	23:07	14.2	0.225	NW 4.8
111	03/16/64	21:25	14.4	0.225	NW 1.6
112	03/20/64	21:11	14.4	0.225	NW 1.6
113	04/07/64	22:11	14.0	0.225	NW 1.6
201	05/26/64	23:41	14.6	0.025	NW 1.6

(Continued)

Test	Date	Time, LST	Amount Released Air, kg	Ground, kg	Direction/ Distance, km
202	05/30/64	23:28	14.0	0.025	NW 4.8
203	06/10/64	00:05	13.3	0.025	NW 1.6
204	06/13/64	01:07	13.6	0.025	NW 1.6
205	06/28/64	23:55	14.0	0.025	NW 1.6
206	07/06/64	00:55	13.8	0.025	SE 1.6
207	07/07/64	22:53	14.9	0.025	NW 1.6
208	07/15/64	23:44	14.7	0.025	SE 1.6
209	07/28/64	22:50	14.1	0.025	NW 1.6
210	08/02/64	22:45	14.0	0.025	NW 1.6
211	08/07/64	01:24	10.4	0.025	NW 1.6
212	08/07/64	22:31	14.1	0.025	NW 1.6
All		332	2.6		

Communities affected:
The nearest permanent communities potentially affected were the town of Cass Lake, approximately 8 km to the northwest of the test area, and Bemidji, 32 km northwest. These communities could only have been affected by Tests 103, 206, and 208, when the wind was blowing in their direction.

On current maps, there are now also campgrounds (Norway Beach Campground and Ojibway Campground) located on the shore of Cass Lake approximately 2 km closer than the community of Cass Lake.

To the southeast, the nearest downwind communities on current maps are Federal Dam (approximately 24 km) and Boy River (approximately 32 km).

Distances from releases to affected communities:
Cass Lake is approximately 8 km from the center of the test array, and so approximately 10.6 km from the release point for the three releases that might have affected it.

Federal Dam is approximately 27 km from the nearest releases that could have affected it.

Weather conditions:
Releases were always performed under stable or neutral meteorologic conditions, with no rain present or forecast, and with the wind direction close to the axis of the test array and expected to remain that way during the period of the test (1 to 3 h). Wind speeds during winter were 0.7 to 2.6 m/s, and during summer 0.2 to 0.7 m/s, at 10 m altitude.

Other materials released:
No other materials are mentioned.

Maximum concentrations and time-integrated concentrations:
Air and ground releases could be distinguished by the different colors of fluorescence of the particles. Only concentrations from the aerial releases were evaluated. The mass released at ground level was much smaller, and ground-released particles did not travel extensively outside the measurement area, which was unpopulated except for a crew of 30 to 40. Within a few meters of the ground-level bomblets, local instantaneous concentrations could be very high (approximately 6 g/m^3 at the instant of burst and within the 1-m radius initial spherical cloud).

The following table indicates the highest 15-min average concentration, estimated by the original authors, together with measured exposures. Individual measurements are probably accurate only to within a factor of 2. (There was a standard deviation of a factor of 1.8 in the relative measurements of co-mounted Rotorods and sequential samplers.)

Test	Max. 15-min Conc., µg/m³	Max. Integrated Conc., µg-min/m	Test	Max. 15-min Conc., µg/m³	Max. Integrated Conc., µg-min/m³
102	2.4	52	201	1.6	59
103	9.1	16	202	0.6	133
104	2.2	68	203	6.2	178
105	2.2	77	204	3.3	104
106	5.5	34	205	3.6	1408
107	2.9	94	206	2[a]	40
108	5.6	17	207	2.6	4410
109	1.5	35	208	4.2	80
110	1.8	63	209	2.0	54
111	1.1	34	210	3.6	153
112	1.8	83	211	3.3	46
113	5.5	49	212	9.1[b]	77
All	9.1	2779			

[a]Estimated. The original authors did not make any estimate for this experiment.
[b]This was reported as the maximum by the authors, although it conflicts with the measured exposure (a 15-min average concentration of 9.1 µg/m³ itself contributes 136.5 µg-min/m³ to exposure). The authors apparently interpolated to estimate the maximum concentrations, while all the exposures are measured.

Maximum concentrations and time-integrated concentrations in any populated area:

Air and ground releases could be distinguished by the different colors of fluorescence of the particles. Only concentrations from the aerial releases were evaluated. The mass released at ground level was much smaller, and ground-released particles did not travel extensively outside the measurement area, which was unpopulated except for a crew of 30 to 40.

The following table gives estimates of the concentration reached in the nearest populated area. These were obtained by using standard gaussian plume dispersion models, together with estimates of the relevant meteorologic parameters. They should not be considered more accurate than a factor of 3, when interpreted as concentrations that might have affected these populated areas—actual plume travel was not measured.

Test	Max. Conc., µg/m³	Max. Integrated Conc., µg-min/m³	Area Affected
102	0.3	10	Federal Dam
103	3	50	Cass Lake
104	0.4	10	Federal Dam
105	0.4	18	Federal Dam
106	0.9	50	Federal Dam
107	0.4	25	Federal Dam
108	0.4	30	Federal Dam
109	0.4	30	Federal Dam
110	0.4	30	Federal Dam
111	0.5	12	Federal Dam
112	0.4	30	Federal Dam
113	1	70	Federal Dam
201	1	140	Federal Dam
202	0.9	110	Federal Dam
203	1.5	200	Federal Dam
204	0.5	70	Federal Dam
205	0.9	190	Federal Dam
206	1.5	80	Cass Lake
207	2.0	300	Federal Dam
208	2.6	160	Cass Lake
209	1.5	110	Federal Dam
210	2.0	100	Federal Dam
211	0.4	30	Federal Dam
212	0.9	55	Federal Dam
All	3	290	Cass Lake
All	2	1620	Federal Dam

Other comments:

Table 3-8 in the original, from which the exposure estimates above have been taken, is incorrectly characterized in the report. It is said to be in particle-minutes per liter, but actually corresponds to 1/15 of the exposure in particle-minutes per liter, and uses a nominal calibration factor for the Rotorods in place of the measured calibration factor.

APPENDIX B 203

A crew of approximately 30 local people were employed to operate the sampling grid during testing.

References to other possible FP studies that might be relevant.

1-6 Geophysics Corp. of America. Technical Report 64-3-g. Meteorological Prediction Techniques and Data System, by H.E. Cramer et. al. Contract DA42-007-CML-552. Final Report. Mar. 1964. AD444, 197.
Unclear whether this study used FP.

1-7 Bendix Corp. Bendix Systems Division. Jungle Canopy Penetration. Vol. 1. Diffusion Measurements. Contract DA42-007-530. Final Report. Aug. 1961-Jan. 1963 (by Hamilton et. al., according to p. 1-32 of text).
FP was used in this study, but what type is not stated. Sponsored by Dugway.

1-8 Meteorology Research. Report 63-FR-108. Dissemination and Evaluation of a Tracer Material Release (Big Jack) (U), Vol. 1, by T.B. Smith and F. Vukovich. Contract DA42-007-AMC-15(X). Final Report. Nov. 29, 1963. AD 347, 791. Confidential. (Text p. 1-31 indicates authors as Smith and Leavengood.)
Unclear whether this study used FP.

1-9 I.A. Smith and M.E. Singer. Personal conversations (1966). (Text indicates experiments at Brookhaven National Laboratory.)
Unclear whether this study used FP.

DESCRIPTION OF TEST

Name of test: *San Francisco* (no code name identified)

Reference-list number: 20

Reference: J.A. Murray, T.S. Brown, and F.X. Webster. FP Tracer Co-Dispersal and Sampling Studies. Technical Report 135. Metronics Associates, 3201 Porter Drive, Stanford Industrial Park, Palo Alto, Calif. Contract DA-42-007-

AMC-240(R) under RDT&E Project 1V025001A128, Meteorological Aspects of CB Program, for U.S. Army Dugway Proving Ground, Utah. Dec. 1968.

Principal object:
To investigate the effect of co-dispersal of FP with other materials and tracers and to improve the representation of the efficiencies of the co-dispersal dissemination systems and the air sampling systems.

Site selection:
Three test sites were used. No information on site selection is given, but we may surmise that they were selected for their meteorologic characteristics:
1. The Industrial test site was close to large, low buildings and surrounded by vacant, multi-acre building sites.
2. The Bayshore Flats test site was adjacent to San Francisco bay and frequently exposed to steady onshore breezes.
3. The Foothill test site was in the rolling foothills in the eastern slope of the coastal mountains that run the length of the San Francisco Peninsula.

No further geographic information was given on the location of these sites.

Number and nomenclature of releases:
A total of 18 releases is described in seven field experiments. The field experiments were identified by the nomenclature FE 146, FE 151, FE 152, FE 154, FE 155, FE 156, and FE 157. Within each, there were a variable number of releases, each labeled as a "trial" and indicated by alphabetic letters:

FE	Trial	Site
146	D	Industrial
151	A	Industrial
152	A, B, C	Bayshore Flats
154	A, B, C, D	Bayshore Flats
155	A, B, C	Foothill
156	A, B, C, D	Foothill
157	A, B	Bayshore Flats

Test conditions:

These tests were primarily designed to measure efficiencies of equipment, not dispersal characteristics over large areas. At the Industrial test site, the field tests were in the vacant building sites, with the sampling station 100 m downwind of the generator. At the Bayshore site, FP cloud travel distances were greater than 1,000 m. At the Foothill site, samplers were located about 200 m downwind of the generators.

Test material:

Various combinations of ZnCdS FP that fluoresce different colors were co-dispersed and, also in 156 A, B, C, and D, simultaneously dispersed. The following were used in various combinations:

FP Tracer Type	Color	Lot	Particle-size Parameters Particles per g	Mass-mean Diameter, μm	Number-mean Diameter, μm
2267	Yellow	DPG 11	1.46×10^{10}	3.2	—
2267	Yellow	H 511	2.15×10^{10}	2.8	1.9
2267	Yellow	H 443	1.55×10^{10}	3.1	2.1
3206	Green	H 396-1	1.33×10^{10}	3.3	—
3206	Green	H 513	1.54×10^{10}	3.14	2.2
4003	Green	Exptl.[a]	1.5×10^{7}	31.6	25
2220	Orange	H 399	1.45×10^{10}	3.2	—
2205	Blue	H 397	1.23×10^{10}	3.4	—
1491	Red	Exptl.	5×10^{8}	10.1	2.4
4763	Red	Exptl.	3.04×10^{9}	5.4	3.5
4763	Red	Exptl.	2.67×10^{9}	5.6	3.6

[a]Exptl., experimental.

Place of release:
See "Number and nomenclature of release" above. No further information is provided.

Dates, times, and quantities of release:
The times of the tests are not generally given.

FE	Trial	Date	Time, PDT	Quantities of ZnCdS, g
146	D	3/25/64		23.5
151	A	10/18/65		1.0
152	A, B, C	12/1/65		100, 128.3, 162.7
154	A, B, C, D	3/24/68		147.4, 186, 217, 248.5
155	A, B, C	4/20/66		151.2, 171.9, 160.5
156	A	8/4/66	12:14-12:24	441.3
	B[a]	8/4/66	14:21-14:31	354.6
	C[a]	8/4/66	15:04-15:14	396.9
	D	8/4/66	15:53-16:03	356.5
157	A, B	5/23/67		16, 14.9

[a]The mass of the organic FP in FE 156 B and C has been omitted.

By site, the total quantities involved are the following:

Site	Quantity of ZnCdS
Industrial	24.5 g
Bayshore	1.22 kg
Foothill	2.03 kg

Communities affected:
Unknown. There is insufficient detail to locate the test sites beyond the description given above in "Site selection."

Distances from releases to affected communities:
Unknown. There is insufficient detail to locate the test sites. All were apparently in locally deserted or vacant areas out to the distance of the samplers. The one potentially useful site photo is too indistinct to draw any conclusions.

Weather conditions:

There is no mention of weather conditions, except that wind speeds are given for some experiments:

FE 152   Trial averages 6.5-7.6 mph, extreme range 3-10 mph
FE 154   Trial averages 7.2-11.9 mph, extreme range 5-12.5 mph
FE 156   Trial averages 5.6-8.5 mph, extreme range 3.0-11.2 mph
FE 157   Averaged 10-16 mph from north-northwest, extreme range 8-20 mph

Other materials released:

In two releases (156 B and C), there was an additional admixture of an experimental, low-density, organic FP (3.5% and 3.3% by weight of the total releases, respectively). In four releases (FE 156 A, B, C, and D), nonviable BG spores prestained with crystal violet were co-dispersed and sampled with the FP.

Maximum time-integrated concentrations:

For the Industrial site, no measurement data are given, but the masses involved were small compared with the other sites. The sums over all trials of maximum measured time-integrated concentrations were the following:

Location	Cumulative time-integrated concentration, $\mu g\text{-}min/m^3$
Industrial site	(data not available)
Bayshore Flats site	1100
Foothill site	1900

Maximum time-integrated concentrations in any populated area:

Not known. Probably no higher than the maximum time-integrated concentrations just given, because the tests apparently were carried out in vacant areas, with no people exposed between the generation point and the point of measurement.

Maximum concentrations:
For the Industrial site, no measurement data are given, but the masses involved were small compared with the other sites. For the Bayshore tests, only exposure (not concentration) information is available; but concentrations were probably similar to the Foothill site at similar distances (at the Bayshore site, the samplers were located further away from the generators). At the Foothill site, measurements were made with drum pulsed samplers, giving measurements of 6-s and 30-s average concentrations, in addition to the dosage information given by membrane filters and Rotorods.

Location	Maximum measured concentration $\mu g/m^3$
Industrial site	(not measured)
Bayshore Flats site	(not measured)
Foothill site	170

The 6-s sampling clearly shows the effects of plume waver, even at 200 m from the dispersal point—during 10 min of continuous dispersion, several peaks in concentration are observed in the record.

Maximum concentrations in any populated area:
Unknown. Unlikely to be higher than the maximum concentration given above, because the trials were apparently carried out in vacant areas.

Other comments:
This paper includes an analysis of some results from a previous field trial (FE 146, Trial D, 25 Mar. 1964). The implication is that there were other prior field trials.

These trials were small-scale calibration and feasibility trials, rather than dispersion tests.

Also mentioned in the paper are aerosolization tests designed to evaluate the separation of the mixture of FP particles of different colors during aerosolization. No details of the methods used in these tests are given, and it is not clear whether they involved outside dispersion.

DESCRIPTION OF TEST

Name of test: *Urban Diffusion Project*

Reference-list number: 22

Reference: G.R. Hilst and N.E. Bowne. A Study of the Diffusion of Aerosols Released from Aerial Line Sources Upwind of an Urban Complex. Vol. 1: Final Report; Vol. 2: Final Report: Data Supplement. The Travellers Research Center, 250 Constitution Plaza, Hartford, Conn. Contract DA-42-007-AMC-37(R) under RDT&E Project 1V025001A128, Meteorological Aspects of CB Program, for U.S. Army Dugway Proving Ground, Salt Lake City, Utah. July 1966.

Principal object:
"To determine the manner in which an isolated city affects the structure of atmospheric motions, and thereby the manner in which airborne materials will be diffused and transported as they move over and through the city."

Site selection:
A city was required that was isolated, located on flat land away from large water bodies, with distinct residential, park, industrial, and commercial areas, and large enough to affect dispersion processes but not so large as to preclude reasonable operational costs. The last was most difficult to quantify. A city diameter of about 10 miles was selected, and 12 cities (populations of 150,000 to 300,000) examined in the plain areas of Kansas, Illinois, Indiana, Ohio, and Michigan. Fort Wayne, Ind., was selected as meeting all requirements and also providing "several tall towers

properly located around and within the city, and a cooperative and genial citizenry and town government."

Number and nomenclature of releases:
A total of 75 separate releases in 23 experimental periods. Each experiment (up to four releases) was labeled as TRC 64-1 through TRC 64-6, and TRC 65-1 through TRC 65-18. The number TRC 65-4 was assigned, but that experiment was canceled. Within each experiment, four releases were planned, labeled Y1, Y2, G1, and G2, representing two releases each of yellow- and green-fluorescing FP.

Test conditions:
Releases were from two Piper Apache aircraft flying at less than 240 m altitude perpendicular to the wind direction, upwind of the city, over a 20-mile release line, with a release rate of 2.5 lb/mile. During each of two release runs, each aircraft released a differently fluorescing tracer. Surface samplers were located on five crosswind lines (modified by street location requirements), one at the upwind edge of the city, three dividing the city into its major land-use variations, and one on the downwind edge. Each line contained 50 samplers at roughly 0.5-mile intervals, and extended out from the city into the rural areas to the northeast, to allow simultaneous measurements over city and rural areas. The sampling network limited allowable wind directions to the northeast or southwest. In addition, 25 Millipore filter samplers were distributed more or less at random. Upwind and downwind vertical distributions of exposure were measured using samplers strung up a balloon tether and a guy wire of the 244-m WANE-TV tower. While the experiment was planned to allow releases with the wind from the northwest or the southeast, all experiments were carried out with a generally northwest airflow, although there was a substantial variation of direction, with the airflow almost parallel to the sampling lines in several experiments (so the plume almost missed the sampling array).

Test material:
Two ZnCdS tracers that fluoresce different colors were used (one released by each aircraft).

APPENDIX B

FP Tracer Type	Color	Lot	Particle-size parameters	
			Particles per g	Mass-mean diameter, μm
2267	Yellow	H-395	1.32×10^{10}	3.31
3206	Green	H-396	1.45×10^{10}	3.2
3206	Green	H-391[a]	1.62×10^{10}	3.09

[a]Used in one release (unspecified) in 1964.

Place of release:
In all cases, slightly northwest (upwind) of Fort Wayne, Ind., along a roughly 20-mile track approximately northeast-southwest, perpendicular to the wind direction (see below for distances).

Dates, times, and quantities of release:
The following table shows experiment number and date. Within each experiment, four releases (labeled Src. Y1, Y2, G1, G2) were planned. The table shows the mass released ("?" if this is not given), and the length, altitude, and upwind distance of the release path (distance upwind from the nearest sampling grid line on the upwind edge of the city). Also included are reasons specified for nonrelease. TRC 64-1 was said to have few analyzable data, and no further information was provided on these releases. (It is not specified whether all four releases were made.)

Name	Date	Time, LST	Src	Total Amount, kg	Relative Length, m	Relative Altitude, m	Upwind Distance, m
TRC 64-1	02/02/64		Y1	?	No further information available		
			Y2	?	No further information available		
			G1	?	No further information available		
			G2	?	No further information available		
TRC 64-2	03/22/64	00:28	Y1	22.7	20,985	122	1600
			Y2		No reason given		
		00:38	G1	22.7	27,725	122	4800
			G2		No reason given		
TRC 64-3	07/25/64		Y1	?	One release of yellow tracer		
			Y2		Unfavorable winds		
			G1		Unfavorable winds		
			G2		Unfavorable winds		

(Continued)

Name	Date	Time, LST	Src	Total Amount, kg	Relative Length, m	Relative Altitude, m	Upwind Distance, m
TRC 64-4	07/29/64	20:45	Y1	20.8	20,124	122	1600
		21:30	Y2	21.3	20,275	122	1600
		20:52	G1	22.7	21,230	183	1600
		21:35	G2	22.5	22,075	183	1600
TRC 64-5	08/09/64	21:57	Y1	16.4	22,470	126	1600
			Y2		Aircraft out of commission		
			G1		Aircraft out of commission		
		22:32	G2	21.7	21,490	126	1600
TRC 64-6	08/13/64	20:05	Y1	18.6	22,946	183	1600
		21:29	Y2	17.9	24,019	183	1600
		20:10	G1	20.8	20,620	91	3200
		21:34	G2	22.0	21,156	91	3200
TRC 65-1	10/24/65	20:02	Y1	22.7	15,704	214	1600
		21:20	Y2	22.7	18,420	244	1600
		20:18	G1	22.7	28,807	92	1600
		21:35	G2	22.7	35,586	122	1600
TRC 65-2	10/26/65	19:00	Y1	22.3	26,333	183	1600
		21:01	Y2	22.8	26,909	183	1600
		19:15	G1	22.7	28,539	91	1600

APPENDIX B 213

Name	Date	Time, LST	Src	Total Amount, kg	Relative Length, m	Relative Altitude, m	Upwind Distance, m
(Continued)							
		21:18	G2	22.8	27,661	91	1600
TRC 65-3	11/08/65	19:07	Y1	22.5	26,869	183	3200
		20:34	Y2	22.6	26,869	183	3200
		19:22	G1	23.2	35,513	91	3200
		20:49	G2	22.6	32,924	91	3200
TRC 65-5	11/16/65	20:03	Y1	22.0	26,903	91	1600
			Y2		Severe turbulence		
		20:12	G1	22.6	33,213	91	4800
			G2		Severe turbulence		
TRC 65-6	11/17/65		Y1		Dispenser malfunction		
		20:25	Y2	22.1	23,992	183	1600
		18:45	G1	22.5	32,133	91	1600
		20:47	G2	22.2	26,387	91	1600
TRC 65-7	11/29/65	18:42	Y1	22.5	26,333	183	3200
		20:44	Y2	22.8	28,203	183	3200
		19:00	G1	22.7	33,400	91	3200
		21:00	G2	22.8	30,873	91	3200
TRC 65-8	12/05/65	18:30	Y1	22.8	24,536	183	4800
		20:00	Y2	22.6	24,791	183	4800
		18:45	G1	22.7	15,751	91	4800
		20:15	G2	22.7	30,041	91	4800
TRC 65-9	12/06/65	18:00	Y1	22.9	25,823	214	3200
		19:45	Y2	23.0	26,514	214	3200
		18:15	G1	22.9	33,977	122	3200
		20:00	G2	23.0	30,464	122	3200
TRC 65-10	01/06/66	19:03	Y1	22.2	23,892	122	3200
			Y2		Cloud cover, release canceled		
		19:08	G1	22.4	28,371	122	3200
			G2		Cloud cover, release canceled		
TRC 65-11	01/10/66	19:04	Y1	22.3	27,452	122	4800
		20:45	Y2	22.5	29,116	122	4800
		19:09	G1	22.2	25,602	122	4800
		20:50	G2	22.8	28,056	122	4800
TRC 65-12	01/13/66	19:00	Y1	22.2	24,831	122	3200
		20:30	Y2	24.3	24,844	122	3200
		19:05	G1	22.4	26,661	122	3200
		20:35	G2	22.4	26,742	122	3200
TRC 65-13	01/23/66	18:30	Y1	22.6	24,784	91	3200
			Y2	?	Malfunction, dispenser jammed		

Name Date	Time, LST	Src	Total Amount, kg	Relative Length, m	Relative Altitude, m	Upwind Distance, m
(Continued)						
	18:35	G1	20.6	27,493	91	3200
	20:20	G2	11.3	16,000	91	3200
TRC 65-14 01/27/66	19:45	Y1	24.5	27,790	122	1600
	21:15	Y2	22.9	26,266	122	1600
	19:51	G1	22.7	28,472	122	1600
	21:21	G2	22.7	21,411	122	1600
TRC 65-15 01/28/66	19:00	Y1	22.5	24,462	91	1600
		Y2		No reason given		
	19:05	G1	22.7	28,968	91	1600
		G2		No reason given		
TRC 65-16 01/29/66	19:00	Y1	22.5	25,240	91	1600
	20:15	Y2	22.5	26,580	91	1600
	19:04	G1	23.1	28,834	91	1600
	20:20	G2	22.4	25,696	91	1600
TRC 65-17 02/04/66	18:43	Y1	22.5	26,695	91	1600
		Y2		No release		
	18:48	G1	22.6	27,399	91	1600
	20:08	G2	22.5	24,905	91	1600
TRC 65-18 02/04/66	23:50	Y1	26.9	22,509	122	1600
		Y2		No reason given		
	23:55	G1	22.5	19,526	122	1600
		G2		No reason given		

The total mass released was approximately 1,650 kg, when account is taken of the (up to) five releases for which no measured quantity is given.

Communities affected:
The principal community affected was Fort Wayne, Ind., together with other any rural communities within approximately 10 miles to the northeast.

Distances from releases to affected communities:
The releases were 1.6 to 4.8 km northwest (upwind) of the edge of the city of Fort Wayne. Some people would have been closer to the releases.

Weather conditions:
The intention was to release tracer into a stable atmosphere or a neutral atmosphere with inversion conditions. Most releases were performed in clear weather, but there were traces of snow during some. The atmosphere was generally clear, with visibility exceeding 10 miles.

Other materials released:
No other releases are mentioned.

Maximum time-integrated concentrations and maximum time-integrated concentrations in any populated area:
For this set of experiments, these two concentrations are synonymous. Maximum concentrations were measured on the sample line nearest the releases, so that slightly higher concentrations might have occurred closer to the release lines.

The cumulative maximum observed integrated concentration was approximately 410 µg-min/m^3. This estimate is an average over 10 adjacent sample points and so might be slightly low compared with the single highest sample point.

Maximum concentrations and maximum concentrations in any populated area:
For this set of experiments, those two measurements are synonymous. No direct (i.e., short-term) measurements of concentration were made. A visual scan of the approximately 16,250 exposure measurements indicated just four that exceeded 1,000 particle-min/L (about 70 µg-min/m^3). The largest, 1,878 particle-min/L (about 130 µg-min/m^3), occurred during TRC 65-18, release Y1, at one end of the sampling grid. Dispersion modeling, roughly matching modeled and measured dispersion parameters, indicates peak concentrations around 10-20 µg/m^3.

Other comments:
Too many data to evaluate all points separately. Summary averages over 10 sample points (as presented in the original report) have been used.

DESCRIPTION OF TEST

Name of test: *Oceanside Shoreline Diffusion Program*

Reference-list number: 24

Reference: T.B. Smith, and B.L. Niemann. Shoreline Diffusion Program, Oceanside, Calif.. Report MRI 169 FR-860. Meteorology Research, 464 W. Woodbury Road, Altadena, Calif. Contract DA 42-007-AMC-180(R) for the Deseret Test Center, Fort Douglas, Utah. Nov. 1969.

Principal object:
To understand meteorologic processes from a short distance offshore to 15 km inland, particularly in the transition zone near the shoreline, and determine the effects of such processes on diffusion from various types of sources.

Site selection:
The site was required to have frequent onshore winds and moderate changes in terrain roughness near the coastline. Secondary requirements were for a suitable surface road network, and minimal interference with air traffic. Sites examined included Eglin Air Force Base in Florida; Hoquiam, Washington; Coos Bay, Oregon; the vicinity of Bodega Bay in northern California; and the coastline immediately south of Oceanside, California. The site selected is located between Oceanside and Del Mar along the southern California coast, with Palomar airport in the middle of the test area. Selection of this site was primarily on the basis of inland terrain.

Number and nomenclature of releases:
Offshore, there were 10 elevated line-source trials, five surface line-source trials, and five surface point-source trials. Offshore elevated line and surface source trials proceeded concurrently with FP that fluoresced different colors, and were numbered 27-32 and 35-38. Thirty-five point-source trials were conducted from onshore locations; 14 consisted of two

APPENDIX B

simultaneous point sources. Onshore trials were numbered 1-26, 33-34, 51-55, and 61-62.

Test conditions:
Elevated line releases were from a C-47 aircraft flying along a 40-km north-northwest and south-southeast track passing approximately 2.25 miles offshore at Leucadia and Encinitas, and extending from north of Oceanside to south of Cardiff-by-the-Sea. FP was released at 1,500 g/min over a 10-min release interval. Boat releases were from a fishing boat, with the line releases on the same track as the aircraft releases, but taking about 2 h to traverse the track. The releases were timed so that the aircraft passed the boat at the center of the release line, and the release rate was designed so that about the same quantity was released from boat and aircraft. Point releases from the boat were at variable places along the release track of the aircraft to obtain maximum coverage by the samplers. Release rate was 110 g/min for 5 min.

Onshore point-source releases were from locations near San Marcos, Carlsbad, and La Costa, and took the form of instantaneous bursts using a compressed-air-driven disseminator, 50 g per burst. All Carlsbad and La Costa releases involved simultaneous releases of 50 g each of yellow and green FP, from locations spaced by 100 m.

Sampling was by an onshore network at about 140 locations along major highways and extending 20 km along the shore between Carlsbad and Cardiff-by-the-Sea, and 21 km inland to Interstate 15. Sampling stations were principally individual Rotorod samplers, supplemented by other special sampling systems including Rotorod equipped towers (up to 120 m height), sequential samplers, vertical sampler arrays (2, 4, 6, and 8 m), 100-m square grids of samplers, five dense sampling lines, vertical arrays 50 m downwind of the onshore point sources, and aircraft samplers.

Test material:
ZnCdS: Green fluorescing, Lot H-396, mass mean diameter 3.2 µm, 1.45×10^{10} ppg; Yellow fluorescing, Lot H-395, mass mean diameter 3.31 µm, 1.32×10^{10} ppg.

Measured average dissemination efficiency was approximately 67% (aircraft line source), 85-88% (boat sources), and 26.5% (onshore point sources).

Place of release:
The line releases were along a 40-km north-northwest and south-southeast line offshore, traversing approximately 2.25 miles off Leucadia and Encinitas. By implication (no adequate documentation), the line was approximately centered on Leucadia, which was roughly in the center of the shoreline sampling line. Point boat releases were at locations on this release line, selected so that the plume would traverse the most dense arrays of samplers. Onshore releases of much smaller quantities of material were from sites near Carlsbad (apparently within a few hundred feet of areas that on current maps show local streets), San Marcos (not located), and La Costa (apparently more than 0.5 mile from inhabited areas).

Dates, times, and quantities of release:
Elevated line-source trials (yellow FP):

Trial	Date	Time, PDT hr:min	Duration, min:s	Altitude, m	Amount, kg
27	07/06/67	17:51	11:00	61	14.62
28	07/07/67	18:59	10:30	122	16.09
29	07/08/67	19:01	10:00	61	12.41
30	07/09/67	16:59	10:25	61	17.78
31	07/11/67	15:00	11:00	61	16.73
32	07/12/67	17:58	09:45	61	15.87
35	07/14/67	18:59	09:50	61	16.19
36	07/15/67	17:00	10:43	61	16.24
37	07/16/67	15:00	09:55	122	16.1
38	07/17/67	16:59	10:17	152	16.93

Boat source trials (green FP):

Trial	Date	Time, PDT hr:min	Duration, hr:min:s	Type	Amount, kg
27	07/06/67	17:40	05:00	Point	0.554
28	07/07/67	19:04	05:00	Point	0.472
29	07/08/67	19:04	05:00	Point	0.596
30	07/09/67	17:03	05:00	Point	0.54
31	07/11/67	15:03	05:00	Point	0.611
32	07/12/67	17:06	02:08:20	Line	14.87
35	07/14/67	18:01	02:12:00	Line	15.93
36	07/15/67	16:00	02:14:05	Line	16.49
37	07/16/67	14:00	01:54:30	Line	12.06
38	07/17/67	16:00	01:58:30	Line	13.4

Onshore point-source trials:

Trial	Date	Time, PDT hr:min	Location	Color	Amount, kg
1	06/23/67	16:00	San Marcos	Y	0.05
2	06/23/67	16:30	San Marcos	G	0.05
3	06/23/67	17:35	San Marcos	Y	0.05
4	06/23/67	18:21	San Marcos	G	0.05
5	06/24/67	18:30	San Marcos	Y	0.05
6	06/24/67	18:59	San Marcos	G	0.05
7	06/24/67	20:00	San Marcos	Y	0.05
8	06/24/67	20:20	San Marcos	G	0.05
9	06/26/67	18:05	San Marcos	Y	0.05
10	06/26/67	18:45	San Marcos	G	0.05
11	06/26/67	19:46	San Marcos	Y	0.05
12	06/26/67	20:15	San Marcos	G	0.05
13	06/27/67	18:30	Carlsbad	YG	0.1
14	06/27/67	20:03	Carlsbad	YG	0.1
15	06/28/67	18:30	Carlsbad	YG	0.1
16	06/28/67	20:00	Carlsbad	YG	0.1
17	06/29/67	16:10	Carlsbad	YG	0.1
18	06/29/67	18:00	Carlsbad	YG	0.1
19	06/29/67	19:46	Carlsbad	YG	0.1

Trial	Date	Time, PDT hr:min	Location	Color	Amount, kg
(Continued)					
20	06/30/67	18:00	La Costa	YG	0.1
21	06/30/67	20:00	La Costa	YG	0.1
22	07/01/67	16:14	La Costa	YG	0.1
23	07/01/67	18:00	La Costa	YG	0.1
24	07/01/67	20:10	La Costa	YG	0.1
25	06/02/67	16:15	La Costa	YG	0.1
26	06/02/67	18:01	La Costa	YG	0.1
33	07/13/67	18:30	San Marcos	Y	0.05
34	07/13/67	20:04	San Marcos	G	0.05
61	06/22/67	15:18	San Marcos	Y	0.05
62	06/22/67	18:02	San Marcos	Y	0.05
51	07/01/67	21:23	La Costa	G	0.05
52	07/05/67	15:15	La Costa	Y	0.01
53	07/05/67	15:24	La Costa	G	0.01
54	07/05/67	16:39	La Costa	Y	0.01
55	07/05/67	16:48	La Costa	G	0.01
	Total release				237

Y = yellow fluorescing; G = green fluorescing; YG = both yellow and green fluorescing (equal quantities).

Communities affected:
Oceanside, Carlsbad, Leucadia, Encinitas, Cardiff-by-the-Sea, Del Mar, and communities inland of these.

Distances from releases to affected communities:
The offshore releases were approximately 1 mile from Oceanside, 2 miles from Carlsbad, 2.25 miles from Leucadia, Encinitas, and Cardiff-by-the-Sea, and further from other communities by their distance to the shoreline. The onshore releases were presumably at more than 50 m from inhabited areas (because a vertical sampling array of Rotorods was set up at 50 m). The Carlsbad site appears to be within a few hundred feet of in-

APPENDIX B

habited areas on modern maps. The La Costa site appears to be more than 0.5 mile from the nearest inhabited areas. The San Marcos site was not located.

Weather conditions:
Releases were carried out during periods of onshore sea breezes.

Other materials released:
No other materials are mentioned.

Maximum concentrations and maximum time-integrated concentrations:
These concentrations are essentially the same for unpopulated and populated areas.

Trial	Max. Exp., $\mu g\text{-min}/m^3$	Max Conc., $\mu g/m^3$
Onshore point-source trials		
1	18	10
2	13	7
3	16	5
4	35	26
5	24	15
6	7	3
7	139	102
8	70	22
9	23	14
10	18	6
11	188	114
12	50	20
13	53	47
14	80	282
15	183	245
16	694	884
17	486	1741

18	275	537
19	98	42
20	5	0.5
21	176	82
22	21	22
23	9	10
24	14	6
25	32	11
26	106	108
33	5	1
34	12	3
61	NA	NA
62	5.6	2
51	NA	NA
52	8.8	7
53	1.6	1
54	6.2	9
55	1.9	2

Highest values

La Costa	176	108
San Marcos	188	114
Carlsbad	694	1741

Offshore aircraft-released line-source trials

27	7.5	0.2
28	18.5	0.5
29	13.2	0.1
30	6.6	0.2
31	5.1	0.2
32	6.3	0.1
35	19.8	0.3
36	5.3	0.4
37	3.0	0.2
38	6.4	1.2

Trial	Max. Exp., µg-min/m³	Max Conc., µg/m³
Offshore boat-released point-source trials		
27	2.8	0.37
28	8.0	0.91
29	6.2	0.78
30	6.3	0.99
31	7.8	1.31
Offshore boat-released line-source trials		
32	3.8	0.04
35	15.1	0.3
36	2.9	0.2
37	1.6	0.4
38	2.6	0.3
Cumulative	149	1.3

The document contains no concentration information directly and contains few of the raw data but does report the maximum exposures (time integral of concentration data) given here. The concentration estimates shown here were derived from the exposure values, together with reported estimates for standard deviations of plume sizes or plume speed and duration.

For the offshore sources, the total exposure has been added, because some areas probably were affected by all releases (although the total might be an overestimate for any individual point, because the maximum values cited occurred at different points). The point-source estimates apply for a distance of 1-2 km from the point sources and correspond to very narrow (20 to 100 m wide) and short puffs (generally 20 s to a couple of minutes). It is unlikely that any point beyond about 1 km from the release point was affected by more than one or two such puffs, so just the highest values have been reported. Closer to the release points, the short-term concentrations and exposures would be higher, and the exposures might have been cumulative, but over even smaller areas and for shorter times.

Other comments:
There were over 20,000 dosage samples taken. Only condensed results are given, although it is claimed that the maximum exposures were for individual points.

The Carlsbad onshore point-source location is within a few hundred feet of local streets on current maps. It is not known if buildings are located on those streets, nor how recent any such buildings are. No contemporary maps have been consulted.

The text mentions (but does not give mass or exposure information on) a series of eight FP trials, Aug. 8-10, 1967, from Bolsa Island (drilling rig), 90 km north of Oceanside, 1.3 km offshore (T.B. Smith and K.M. Beesmer. Bolsa Island Meteorological Investigation. MRI Report FR-650 for Bechtel Corp., 58 pp., 1967)

DESCRIPTION OF TEST

Name of test: *Dispersion into and within a Forested Area* (no code name identified)

Reference-list number: 27

Reference: Dispersion of Air Tracers into and Within a Forested Area. College of Forest Resources, University of Washington, Seattle, Wash., for U.S. Army Electronics Command, Atmospheric Sciences Laboratory, Fort Huachuca, Ariz., under Grant DA-AMC-28-043-68-G8. Technical Report ECOM-68-G8-1. Sept. 1969.
   Vol. 1: L.J. Fritschen, C.H. Driver, C. Avery, J. Buffo, R. Edmonds, and R. Kinerson (Objec-

APPENDIX B 225

Vol. 2: L.J. Fritschen, C.H. Driver, C. Avery, J. Buffo, R. Edmonds, R. Kinerson, and P Schless (Tabulated data and dispersion patterns).

Vol. 3: L.J. Fritschen, C.H. Driver, C. Avery, J. Buffo, R. Edmonds, R. Kinerson, and P Schless (Analysis and interpretation).

Principal object:
To determine the general features of mass and momentum transport at a forest border interface and simultaneously study the dispersion of small particulates into and within a forest canopy from a clear-cut area in relation to meteorologic conditions.

Site selection:
No selection method is given. The experimental site was in the Charles Lathrop Pack Demonstration Forest of the University of Washington, at the southern border of Pierce County, Washington, 35 km south of Puyallop, 45 km southeast of Tacoma, and 48 km southeast of Olympia.

Number and nomenclature of releases:
The releases were identified by run numbers (1 through 4) and release numbers within runs. Two series of runs were performed, one on the "large grid" of samplers and the second on the microgrid of samplers.

Test conditions:
The research area covered 240 hectares of forested area up-slope of a 200 × 600 m clear-cut area. A 9 × 7 grid of surface level (1 m) Rotorod samplers, with auxiliary elevated samplers on towers located on some of the grid points, was set up with 30-m spacing approximately 70 m from the clear-cut area. Two 5 × 5 sampler microgrids with a spacing of 15 m were interpolated into the larger grid.

Several release mechanisms were tried. Firecrackers were found inefficient. A particle generator also proved to have retention problems. An atomizer using particle-in-water suspensions clogged and caused clumping. A commercial "Hudson" duster was tried. Finally, a compressed-CO_2 driven bomblet was specially designed.

Test material:
Helecon (U.S. Radium), ZnCdS FPs.

It is not clear whether all these are ZnCdS. At least one other reference (listed as "Unnumbered" in "Miscellaneous Notes") indicates that No. 2210 is ZnS. They have all been counted as ZnCdS here.

For releases on the large grid of samplers:

Run	FP Color	Manuf. Code	Lot
1	Green	2210	H775
	Orange	2220	H629
2	Green	2210	H775
	Orange	2220	H629
	Yellow	2267	H779
3 and 4[a]	Green	3206	H848
	Orange	2220	H917
	Yellow	2267	H779

[a]1% hydrophobic silica was admixed on runs 3 and 4.

For releases on the microgrid of samplers:

Run[a]	FP Color	Manuf. Code	Lot
2	Green	3206	Not specified
3 and 4	Green	2210	

[a]No FP was released in microgrid run 1.

Place of release:
Releases were generally within the sampling grid, or within 30 m of the sampling grid. One release was 80 m from the edge of the grid.

Dates, times, and quantities of release:

Run	Release	Date	Time	Method[a]	Total mass, g
Main grid releases					
1	1	10/15/68	16:42	F	F could not
	2	10/15/68	17:02	F	be estimated
	3	10/16/68	12:55	Dyes only	F could not
	4	10/16/68	13:00	G	be estimated
2	1	05/02/69	16:00	A	83
	2	05/02/69	19:19	A	83
	3	05/03/69	07:59	A	83
	4	05/03/69	09:40	A & D	85
	5	05/03/69	11:52	D	89
	6	05/03/69	13:45	D	89
3	1	07/08/69	05:53	P	90
	2	07/08/69	10:18	P	90
	3	07/08/69	13:52	P	90
	4	07/08/69	16:10	P	90
	5	07/08/69	20:37	P	90
	6	07/08/69	22:20	P	90
	7	08/08/69	09:52	P	90
4	1	08/26/69	00:14	P	90
	2	08/26/69	10:44	P	90
	3	08/26/69	13:14	P	90
	4	08/26/69	09:44	P	90
	5	08/26/69	12:12	P	90
	6	08/26/69	15:45	P	90
Microgrid releases					
1	1	07/31/69	16:20	Spores only	
2	1	08/14/69	15:40	A	0.7
	2	08/14/69	16:35	A	0.9
3	1	08/29/69	11:56	A	0.9
	2	08/29/69	12:41	A	0.9
	3	08/29/69	15:17	A	0.9
4	1	09/04/69	17:30	A	0.9
	2	09/04/69	22:49	A	0.9
	3	09/05/69	09:31	A	0.8
	4	09/05/69	11:41	A	0.9
Total release					1700

[a]Dissemination methods: F = firecracker, G = generator, A = atomizer, D = dry particle sprayer, P = pressure release (see "Test conditions" above).

Communities affected:

None. The releases were in the middle of a forested area, below the top of the canopy, and mostly underneath that canopy (some releases were in the adjacent clear-cut area, at times when the wind was blowing into the sampling area). The plumes were small and dissipated rapidly, sometimes to unmeasurably low exposures within the 240 m of the grid in even the downwind direction.

Distances from releases to affected communities

No communities were apparently affected. The location of the sampling grid is about 0.8 miles from the small community of La Grande and about 2 miles from the outskirts of Eatonville.

Weather conditions:

Releases were carried out in all types of weather conditions. They were located so that the plume was directed into the sampling grid.

Other materials released:

There were some relatively unsuccessful releases using fluorescent dyes (sensitivity too low), and releases of spores of *Fomes annosus* stained in 0.5% "Phloxine" (one release was of unstained spores).

Concentrations:

No concentrations or exposures were estimated for this experiment in view of the difficulty of extracting the information[4] and its low value for this report. Concentrations were very high (some unmeasurably so) near the point-source releases, which were spread within and close to (within 80 m of) the sampling grid. However, the low masses released limited the exposures. No people other than the experimenters would have been affected by the releases.

---

[4]It would require retrieving data spread throughout 214 pages and extracting approximately 3,600 data points (9 × 7 grid, 3 colors, 19 experiments).

APPENDIX B 229

DESCRIPTION OF TEST

Name of test: *Meteorological Analog Test and Evaluation (Mate): First Field Test*

Reference-list number: 28

Reference: J.K. Allison, A.V. Duffield, and J.M. Morton. Meteorological Analog Test and Evaluation: First Field Test Operation. Final Report. Meteorological Research Laboratory, Melpar Division of American Standard, 7700 Arlington Blvd., Falls Church, Va. Contract DA 42-007-AMC-3-39(Y), Task VIII B, for U.S. Army Deseret Test Center, Fort Douglas, Utah. June 1970.

Principal object:
To evaluate the equipment systems to be used in the MATE program, which is "a broad, long-range study in applied meteorology," "to measure diffusion and micrometeorology at typical sites and characterize these sites in terms of CB (primarily of diffusion)," and to begin data collection for the MATE program with observations at the first forest analog site.

Site selection:
Sites had to be at least 1 mile square, relatively level, well-forested with mature deciduous trees, and accessible from Melpar laboratories. Two potential sites in Virginia were located, but were subsequently dismissed from consideration (the first because of public concern over the integrity of a Bald Eagle nesting site, the second when it was discovered that a gas transmission line was being built through it). A more extensive search into Maryland located the selected site in the Green Brier Swamp, 12 miles southwest of Cambridge, Md., and permission for use was obtained from the owners of the various parts of the site.

Number and nomenclature of releases:
There were 115 trials, numbered from 1 to 115. Green FP and yellow FP were released in 111. In four, no FP was released. Some additional un-

numbered "trials" involved other operations connected with the MATE program and did not involve any FP release.

Test conditions:
FP releases were made in the center of sampling grids, of which there were four. The main sampling grid consisted of three circular arrays of Rotorod samplers at 100 m radius (every 10 degrees of arc), 200 m and 400 m radius (every 5 degrees of arc). Additional samplers were located on a 1,000-m square ("XM3") grid around the circular grid. Two subsidiary sampling grids ("white" and "yellow") consisting of 100- and 200-m radius circles were constructed, the first off-set by approximately 500 m east from the center of the main grid, the second offset approximately 1 km from the main grid.

The whole site was forested, with trees of typical height 20 to 30 m, and tree density of 150 to 500 per hectare. The last 10 trials were performed after leaf fall to evaluate dispersion characteristics under such conditions.

Test material:
ZnCdS tracer materials. Green-fluorescing, type 3206, Lot SCM FPG-11. Yellow-fluorescing, type 2267, Lot FPY-12. (Note: The analysis dates given for these materials precede the experiment by 4 years.)

Place of release:
Releases were made close to ground level (1 or 1.5 m height, yellow FP, all trials) and at various heights of 1 m to 28 m (green FP). All releases were at the center of one or another of the sampling grids, in most cases with simultaneous release of yellow and green FP in the center of the main grid, occasionally with different colors in the centers of the different grids.

Appendix B

Dates, times, and quantities of release:

| Trial | Date | Time | Mass Release, g | |
			Yellow	Green
1	08/01/69	16:51	35	11
2	08/04/69	11:22	8	5
3	08/06/69	11:00	5	8
4	08/06/69	12:45	8	9
5	08/07/69	10:44	6.4	7.2
6	08/07/69	12:20	7.6	6.2
7	08/07/69	15:38	11	17.6
8	08/07/69	18:01	21.6	24.9
9	08/08/69	09:20	4.1	7.7
10	08/08/69	10:52	9.3	2.4
11	08/11/69	11:15	5.2	10.2
12	08/11/69	14:45	0.8	8.3
13	08/12/69	14:15	4.8	10
14	08/12/69	15:46	2.7	13.1
15	08/13/69	08:20	11.1	11.5
16	08/13/69	10:05	10	10.7
17	08/14/69	08:55	8.6	10.7
18	08/14/69	10:35	10.6	13.8
19	08/14/69	12:15	10	9.6
20	08/14/69	13:54	8.5	12.9
21	08/18/69	11:40	8.5	12.9
22	08/18/69	15:35	7.5	14
23	08/19/69	11:05	7.3	15.8
24	08/19/69	13:00	11.4	4.3
25	08/21/69	09:35	12.2	7.2
26	08/21/69	11:25	8.4	9.9
27	08/22/69	09:20	7.9	21
28	08/22/69	10:45	8.5	10.2
29	08/25/69	11:00	7.5	10.6
30	08/25/69	12:30	6.5	9.9
31	08/26/69	11:05	11.8	9
32	08/26/69	12:46	10.2	11.7
33	08/26/69	16:35	12	9.7
34	08/27/69	10:01	7.9	8.3
35	08/27/69	11:17	9.1	10.1
36	08/27/69	13:10	8.5	14.7
37	08/27/69	14:30	12.7	4.8
38	08/28/69	12:14	8.1	13.7
39	08/28/69	13:40	10.2	8.1

(Continued)

Trial	Date	Time	Mass Release, g	
			Yellow	Green
40	08/28/69	15:34	10	6.8
41	09/04/69	16:10	10	14.5
42	09/04/69	17:56	9.7	8.1
43	09/05/69	09:05	10.2	8
44	09/05/69	10:43	10	9.4
45	09/08/69	10:49	9.3	10.4
46	09/08/69	12:37	9.5	5.1
47	09/09/69	10:40	10	10
48	09/09/69	11:50	5.6	8.9
49	09/10/69	08:36	8.8	10.3
50	09/10/69	10:17	9.6	6.8
51	09/10/69	12:31	12.9	13.7
52	09/10/69	14:00	10.6	11.2
53	09/10/69	15:56	13.1	8.6
54	09/10/69	17:15	8.5	9.9
55	09/11/69	08:59	5.1	10.7
56	09/16/69	12:35	4.7	12
57	09/16/69	14:35	8.9	13
58	09/16/69	16:30	7.5	10
59	09/16/69	18:00	11	11
60	09/17/69	10:40	8.5	12.7
61	09/17/69	12:10	10.5	9.8
62	09/17/69	14:30	11.6	14.2
63	09/17/69	15:45	8.5	10.1
64	09/17/69	17:45	9.2	9.9
65	09/18/69	16:30	10.1	12.3
66	09/18/69	18:15	9.5	7.9
67	09/19/69	07:05	8.9	18.2
68	09/19/69	08:45	6.1	20.6
69	09/20/69	10:18	6.3	7.2
70	09/22/69	15:55	13.6	21.3
71	09/22/69	17:30	14.2	18.3
72	09/23/69	08:30	12.9	18.5
73	09/23/69	10:00	10	15.6
74	09/23/69	16:15	11.2	20.5
75	09/23/69	18:00	8.7	17.7
76	09/24/69	15:15	10.3	22.5
77	09/24/69	16:30	9.1	19.3
78	09/24/69	18:00	12.6	17.9
79	09/25/69	11:01	20.8	23.3

(Continued)

Trial	Date	Time	Mass Release, g	
			Yellow	Green
80	09/25/69	12:30	10.8	19.5
81	09/29/69	14:15	12.2	20.8
82	09/29/69	15:45	10.1	20.7
83	09/30/69	10:03	11.8	21.8
84	09/30/69	11:45	11.2	20.6
85	10/01/69	10:30	13.2	23.8
86	10/01/69	10:30	11.2	19
87	10/03/69	15:30	9.8	18.4
88	10/03/69	17:00	8.3	19.2
89	10/04/69	11:55	14.9	18.9
90	10/04/69	13:04	8.9	19.4
91	10/06/69	16:30	10.2	19.4
92	10/06/69	18:00	11.6	19.6
93	10/07/69	08:20	10.6	21.5
94	10/07/69	10:00	5.3	21.6
95	10/15/69	16:00	10.4	19
96	10/15/69	15:05	10	19.3
97	10/23/69		0	0
98	10/24/69		0	0
99	10/27/69	15:10	9.3	20.4
100	10/28/69	10:00	6.5	28.5
101	10/28/69	14:50	8.7	25.7
102	10/28/69	16:00	6.2	19
103	10/29/69	15:15	7.7	23.4
104	10/30/69	14:00	11.1	20.4
105	10/06/69	15:17	11.2	18.1
106	10/10/69	15:30	6.4	18.8
107	10/11/69	11:20	10.7	19.8
108	10/13/69		0	0
109	10/19/69		0	0
110	10/20/69	14:55	9.7	22.9
111	10/20/69	15:45	8.6	19.6
112	10/29/69	11:30	9	20.5
113	10/29/69	14:45	9.6	20.7
114	10/29/69	16:30	9.2	21.9
115	10/24/69	11:45	24.4	36.2
Total release				2701

Communities affected:
Apparently none. The nearest communities (Bucktown, Longfield, and Seward, although it is unclear from the available maps whether the first two are communities) were located approximately 1 to 2 miles from the release points.

Distances from releases to affected communities:
See "Communities affected" above.

Weather conditions:
Weather conditions are not discussed.

Other materials released:
No other materials are mentioned.

Concentrations:
No concentration or exposure information is reported in the available document (see "Other comments" below). Some tabular information would allow estimation of exposures for some experiments at some sample locations. However, the summary information indicates that the measured exposure averaged over all analyzable trials was decreasing as the inverse 3.02 power of distance from the release point for near-ground-level releases over the range 100 to 400 m distance. Extrapolating the empirical equation given in the document to a distance of 1 km indicates a maximum integrated concentration from all releases (if the wind had been in the same direction all the time) of less than 42 $\mu g\text{-min}/m^3$ and a peak concentration of around 0.03 $\mu g/m^3$ from the largest release.

Other comments:
This document summarized the test operation, but explicitly did not provide data. Raw data are said to be available on 81 reels of magnetic tape. Conclusions are given from various analyses of the test results (e.g., variation of standard deviations for use in standard plume equations).

APPENDIX B 235

DESCRIPTION OF TEST
Name of test: *Operation GOOF*

Reference-list number: 29

Reference: Comparison of Decay Rates for *Bacillus globigii* and Zinc Cadmium Sulfide. BW 1A-56, Operation "GOOF." DPGR 175. Dugway Proving Ground Report, BW Assessment Directorate, Project Order 0016. 27 Apr. 1956.

Trial Report, BWAL, BW 1A-1, 2 and 3-56. Comparison of Decay Rates for BG and FP. Operation "GOOF." BWALTR 38. 7 Nov. 1955.

Trial Report, BWAL, BW 1A-4 and 5-56. Operation "GOOF." CMLRE-DU-MBW. BWALTR 40. 21 Dec. 1955.

Principal object:
To obtain data to make a reliable comparison of the total decay rates for ZnCdS and *Bacillus globigii* (BG) disseminated simultaneously from statically functioned E124 bomblets.

Site selection:
Dugway West Vertical Grid.

Number and nomenclature of releases:
Five trials, named BW 1A-1-56 through BW 1A-5-56, each involving one release.

Test conditions:
FP was released instantaneously from four E124 bomblets statically detonated at the four corners of a 24-ft square centered on the sampling grid. Each test used a 60-degress sector of the West Vertical Grid, with sampling stations on arcs at 300, 600, 1,200, 1,800, 3,600, 5,280, and 7,920 ft

with 15, 29, 41, 55, 55, 55, and 55 sampling stations, respectively. Filter-type samplers were used.

Test material:
ZnCdS produced by New Jersey Zinc Company in Batch CEP 8000, Lot 2. Particle count was 4.51×10^{10} ppg, according to Ref. 47 (QR SAL 448-4), although no statement of particle count was given in Ref. 29 (see "Miscellaneous Notes"). Release was from E124 bomblets, each containing 40 g of FP. The release efficiency was not stated, and is assumed to be 100%, so that the aerosol produced had a particle count of 4.51×10^{10} ppg.

Place of release:
Dugway West Vertical Grid.

Dates, times, and quantities of release:

Trial	Date	Time	Amount, g	Mean Wind Direction, Deg.	Mean Wind Speed at 2 m, mph
BW 1A-1-56	08/23/55	21:40	160	280	10
BW 1A-2-56	09/13/55	22:36	160	20	8
BW 1A-3-56	09/14/55	01:25	160	20	5
BW 1A-4-56	11/01/55	15:14	160	115	9.5
BW 1A-5-56	11/01/55	16:41	160	130	9
Total			800		

Communities affected:
No attempt has been made to estimate concentrations in populated areas.

Distances from releases to affected communities:
No attempt has been made to estimate concentrations in populated areas.

Weather conditions:
See under "Dates, times, and quantities of release" for wind speed and direction. No other information is available.

Other materials released:
Simultaneous release of BG from E124 bomblets.

Maximum time-integrated concentrations:
The following table gives the maximum exposures measured anywhere on the sampling grid and the maximum exposures at 1.5 miles on the farthest sampling arc. The cumulative estimates are the sums of the (two) values for experiments where the wind was going in approximately the same direction. It would be possible to sum individual sampling point values, but the extra information gained is not worth the effort involved. These estimates assume the dispersion efficiency to be 100%, so that the aerosol particles per gram is 4.51×10^{10}.

Trial	Max. Exp., μg-min/m^3	Max. Exp. at 7920 ft, μg-min/m^3
BW 1A-1-5	2415	92
BW 1A-2-5	8297	353
BW 1A-3-5	6003	56
BW 1A-4-5	2459	4
BW 1A-5-5	2431	28
Cumulative	14,300	409

Maximum time-integrated concentrations in any populated area:
No attempt has been made to estimate concentrations in populated areas.

Maximum concentrations:
No time-resolved data were available to allow estimation of concentration.

Maximum concentrations in any populated area:
No attempt has been made to estimate concentrations in populated areas.

Other comments:
DPGR 175 is said to deal with the first of a series of tests. BWALTR 38 and 40 are the initial field reports of the tests, with DPGR supplying an interpretation.

DESCRIPTION OF TEST

Name of test: *Operation SELTZER* (land trials) and *Operation WHITE-HORSE* (sea-to-land trials)

Reference-list number: 30

Reference: Special Report 193. An Experimental Investigation of Viable Aerosol Travel from Sea to Land. U.S. Army Chemical Corps Biological Laboratories, Camp Detrick, Frederick, Md. 24 Sept. 1953.

Principal object:
Primary: To determine the effect of traverse over the sea upon an aerosol cloud of viable organisms formed at sea level.

Secondary: (1) To demonstrate the use at sea of the XB-14B test fixture for dissemination of an aerosol cloud of organisms, (2) to determine the meteorologic factors affecting the aerosol travel from sea to land, and (3) to indicate the nature of the relation between the inert tracer and the viable organisms in the aerosol.

Site selection:
Land trials (Operation SELTZER) were conducted at Area C at Camp Detrick, Frederick, Md. For the sea-to-land trials (Operation WHITE-HORSE), the site along the coast west of Panama City, Florida, was chosen after an on-the-spot survey of several possible sites in the Eastern United States.

Number and nomenclature of releases:
Operation SELTZER involved six trials, named Trials 1 through 6. FP was released in Trials 3 through 6. In Operation WHITEHORSE, there were 12 trials, 3 of which were considered abortive (with no data presented), although there might have been FP releases in all 12.

Appendix B

Test conditions:
In both operations, releases were carried out using modified Stanford generators. In SELTZER, at area C of Camp Detrick, two sets of arrays of samplers were available, for different wind directions. Each array consisted of lines of membrane filter samplers at 350 yd, 700 yd, and 1,400 yd from the potential release points. The release point was chosen depending on wind direction to ensure that the plume crossed the sampling arrays.

In WHITEHORSE, releases were 770 to 6,000 yd out from the shore, at a location chosen so that the plume center would travel over the center of the sampling array. The sampling array for FP was located along the beach (there were further sampling arrays for the organisms).

Test material:
ZnCdS, Lot C P5378 from the New Jersey Zinc Company, mass median diameter 2.25 μm, particle count 3×10^{10} ppg. ("The number of particles per gram is the object of further investigations.") In SELTZER, 8 g per trial were incorporated into the slurry used. In WHITEHORSE, 900 g were disseminated from a modified Stanford generator separately from the other agents released.

Place of release:
SELTZER: Area C was located between West 7th St. and Opossumtown Pike. On current maps, the area appears to have at least two major roads across it. In any case, it is plausible that people were exposed at distances not much greater than the distance to the farthest sampling line (approximately 1,400 yd).

WHITEHORSE: The sampling line for WHITEHORSE was centered at the current location of the community of Biltmore Beach. Releases were from the sea surface at distances of 770 to 6,000 yd offshore, at locations selected so that the wind would carry the plume directly toward the center of the sampling line.

Dates, times, and quantities of release:
SELTZER:

Trial	Date	Start Time	Length, min	Mass, g	Wind, mph	Wind Direction
1	02/14/53	16:47	14	0	8.7	137
2	02/16/53	16:31	12	0	12	177
3	02/18/53	16:32	8	7.88	9.5	221
4	02/19/53	16:16	14[a]	2.7	3.5	202
5	02/19/53	17:55	8	3.75	3	264[b]
6	02/24/53	16:26	8	7.85	8.8	190
	Total release			22 g		

[a]Release for 4 min on, 7 min off, 3 min on.
[b]Veering steadily, no results obtained.

Length of time for the release, and wind speed and direction are given to assist interpretation of the concentration information given below. No measurements were obtained for Trial 5, when the wind veered steadily.

WHITEHORSE:

Trial	Date	Start Time	Length, min	Mass, g	Distance, yd
1	03/24/53	17:43	13	368	770
2	03/27/53	16:51	13	900	Abortive
3	03/29/53	17:38	31	679	1720
4	03/30/53	17:53	27	900	1570
5	04/01/53	17:55	21	734	2730
6	04/18/53	17:40	22	741	3610
7	04/21/53	15:20	23	900	835
8	04/22/53	15:21	22	900	1160
9	04/22/53	17:47	22	900	6080
10	04/23/53	16:20	19	900	1125
11	04/27/53	18:45	23	900	Abortive
12	05/02/53	18:46	23	900	Abortive
	Total release			9.7 kg	

The distance given is the distance offshore of the release. Length of the release is given to assist in interpretation of the concentration information provided below. Trials were labeled "abortive" when there was considerable wind shift, so that the plume went appreciably astray of the sampling line. No sampling data are presented in such instances.

Communities affected:
SELTZER: Current maps indicate that the area around "Area C" at Camp Detrick are relatively highly urban, although this might not have been true in 1953. The concentrations given below correspond to those that might have been experienced just outside Area C (e.g., across Opossumtown Pike).

WHITEHORSE: Current maps place the community of Biltmore Beach at the center of the sampling line for these trials. Contemporary land use information was not provided. A contemporary photograph shows the sampling line on a deserted beach. The contemporary aerial photograph of the area in the photocopy of the report available is not helpful. The concentrations given below correspond to those measured on the beach, and thus might substantially overestimate the concentrations to which any communities might have been exposed.

If the Biltmore Beach area was not inhabited in 1953, the next nearest downwind communities would probably have been Bayview, Saint Andrew, and Baker, 4 miles or more across the Grand Lagoon and St. Andrew Bay, and the two peninsulas defined by these inlets. The contemporary map also shows a Naval Min Counter-measures Station on the second peninsula (between Grand Lagoon and St. Andrew Bay), possibly indicative of a military community.

Distances from releases to affected communities:
See above under "Place of release" and "Communities affected."

Weather conditions:
SELTZER was performed at approximately 16:00 each day, at a period when the atmospheric temperature gradient was reaching neutral. The atmosphere was thus generally stable, with inversions just starting to develop in several of the trials.

The WHITEHORSE trials were supposed to be run under nearly neutral temperature gradients on land. Large recoveries in the early trials and unfavorable winds resulted in a modification of plans, so Trials 6 through 10 were performed under lapse temperature-gradient conditions over land. During Trials 3, 4, 7 and 9, there was a lapse temperature gradient over the sea.

Other materials released:
SELTZER: In these land trials, the FP were incorporated into a slurry that included *Serratia marcescens* (SM), strain 8UK (a simulant for vegetative BW pathogens), and *Bacillus globigii* (BG) (a simulant for spore-forming BW pathogens).

Operation WHITEHORSE: The same slurry of SM and BG as used in SELTZER (but omitting the FP) was co-released with the FP, using separate Stanford generators.

Maximum time-integrated concentrations and maximum concentrations:
These have been estimated by assuming an effective dispersion of 5×10^9 ppg for SELTZER, and 1×10^{10} ppg for WHITEHORSE. The dispersion in a slurry in SELTZER might make this an underestimate, but there is no calibration information available.

SELTZER:
Measurements of integrated exposure (exposure) were made at 350 yd and 1,400 yd from the release point. Populated areas are assumed to have lain just beyond the 1,400 yd mark, so these measurements are relevant for populated areas. The concentration estimate was obtained by dividing the exposure by the period of the release. No measurements were taken in Trial 5, because of veering winds. This trial is thus unlikely to have contributed to a cumulative exposure at the worst case point.

APPENDIX B 243

	At 350 yd		Populated Areas	
	Max. Exp.,	Max. Conc.,	Max. Exp.,	Max. Conc.,
Trial[a]	μg-min/m^3	μg/m^3	μg-min/m^3	μg/m^3
3	740	93	68	9
4	165	24	12	2
5	NA	NA	NA	NA
6	620	78	21	3
Total	785	93	71	9

[a]Trials 1 and 2: no FP release.

WHITEHORSE:

Measurements were taken at the shoreline. If that part of the coast was populated in 1953, then these exposures correspond roughly to population exposures also. No measurements were taken during the abortive trials, but they were abortive because of wind shifts. The cumulative effect at the worst case points, as reported under "Total," is thus unlikely to have been affected. Concentrations were estimated by dividing exposures by the period during which the plume was likely to have been present. These plumes were fairly narrow (of the order of 60-80 yd at the beach), and so few people could have been affected.

Trial	Max. Exp., μg-min/m^3	Max. Conc., μg/m^3
1	6.4E+03	489
2	NA	NA
3	4.5E+03	143
4	2.2E+03	78
5	5.3E+04	2470
6	7.0E+03	315
7	7.3E+03	313
8	4.6E+04	2037
9	1.6E+02	7
10	9.3E+04	4778
11	NA	NA
12	NA	NA
Total	1.5E+05	4778

DESCRIPTION OF TEST

Name of test: *Single Round, E-120 Developmental Studies*

Reference-list number: 31

Reference: Trial Record 230, BW 398-A, Trials 8, 9, 10, and 11. Test Design and Analysis Office, Technical Operations Directorate, U.S. Army Chemical Corps Research and Development Command, U.S. Army Chemical Corps Proving Ground, Dugway, Utah. May, 1958.

Principal object:
Not specified.

Site selection:
Dugway West Vertical Grid.

Number and nomenclature of releases:
Four trials, numbered A-8, A-9, A-10, A-11.

Test conditions:
FP disseminated from two Skil blowers located 10 ft behind and 5 ft sideways from the center of the test grid, relative to the prevailing wind direction. No dissemination time or other information provided. No information on samplers is provided, but the West Vertical Grid uses filter-type samplers in other tests.

Test material:
FP (not otherwise specified).

Place of release:
Dugway West Vertical Grid.

Appendix B

Dates, times, and quantities of release:

Trial	Date	Time	Quantity, g	Wind Direction, Deg.	Wind Speed, mph
A-8	04/03/58	02:51	8.15	139	12.2
A-9	04/10/58	04:54	14.1	28	4.1
A-10	04/18/58	05:09	14.66	9	11.4
A-11	04/22/58	04:51	13.43	305	17
Total			50.34		

Communities affected:
No attempt has been made to estimate concentrations in populated areas.

Distances from releases to affected communities:
No attempt has been made to estimate concentrations in populated areas.

Weather conditions:
See under "Dates, times, and quantities of release" above for wind direction and speed. Relative humidity varied from 54% to 73%, and the temperature gradient over 0.5-4 m varied from $-0.2°F$ to $2.8°F$.

Other materials released:
BG and UL were released from E-120 bomblets, with dissemination periods of 25 s.

Maximum time-integrated concentrations:
No data are given.

Maximum time-integrated concentrations in any populated area:
No data are given.

Maximum concentrations:
No data are given.

Maximum concentrations in any populated area:
No data are given.

Other comments:
This short report appears to be from a time when FP use is so routine that no particle counts are given in the data report if they are not a principal object of the exercise.

DESCRIPTION OF TEST

Name of test: *Check Test of West Vertical Grid*

Reference-list number: 32

Reference: A.T. Hereim and J.E. Frese. Check Test of West Vertical Grid, Dugway Proving Ground, Utah. Summary Report. RDTE Project 1-X-6-65704-D-634-06. USATECOM Project 5-CO-413-000-013. DTC Project DTC B-008. Deseret Test Center, Fort Douglas, Utah. Oct. 1970.

Principal object:
To check the mechanical performance of the newly rebuilt West Vertical Grid, to examine behavior of aerosols on the grid, and to provide operating experience with the equipment.

Site selection:
Dugway West Vertical Grid.

Number and nomenclature of releases:
Six trials, numbered 1 through 6.

Test conditions:
A Mark IX disseminator was used to dispense approximately 35 g of FP almost instantaneously for each trial. Membrane filters were placed at each station on the vertical portion of the grid, and operated for 2 min about the firing time of the Mark IX.

Test material:
FP, particle count of 4.16×10^{10} ppg.

Place of release:
West Vertical Grid, Dugway Proving Ground.

Dates, times, and quantities of release:
During Feb. and Mar. 1970, release of 35 g per test, total 210 g. No further information is available.

Communities affected:
No attempt has been made to estimate concentrations in populated areas.

Distances from releases to affected communities:
No attempt has been made to estimate concentrations in populated areas.

Weather conditions:
Trials 1-3 were carried out in the daytime (lapse conditions). Trials 4-6 were carried out at night (inversion conditions).

Other materials released:
"Only fluorescent ZnCdS particles (FP) were released as an aerosol because of constraints on open air testing."

Maximum time-integrated concentrations:
The following maximum exposures (which have been adjusted for dissemination efficiency) were recorded at 1 or 2 m height on the vertical

grid, 27 m from the center of the grid. (There is no statement that the Mark IX disseminator was placed at the grid center, although that is the natural presumption, and photos seem to indicate it.)

Trial	Exp., µg-min/m^3
1	24,402
2	29,887
3	23,166
4	30,218
5	16,539
6	12,864
Cumulative	<137,000

The cumulative total given here as an upper bound is the sum of the maximum exposures listed. Because the West Vertical Grid was rotated to align it with the wind and no details were provided on wind directions, an accurate cumulative value cannot be obtained.

Maximum time-integrated concentrations in any populated area:
No attempt has been made to estimate concentrations in populated areas.

Maximum concentrations:
No time-resolved data were available to allow estimation of concentration. These data were collected in 1 min, and it seems likely that the total exposure time this close to the release was of the order of 10 s.

Maximum concentrations in any populated area:
No attempt has been made to estimate concentrations in populated areas.

Other comments:
Reference is made to the "T3-665" trials, which were FP dissemination efficiency trials for the Mark IX disseminator, using FP Lot H-387 and others. Reference: Dugway Proving Ground, Utah, Final Report, Technology Test of FP Fluidizers, by W.A. Brown and J.E. Frese, RDT&E Project 1V025001A128, June 1968.

APPENDIX B

DESCRIPTION OF TEST

Name of test: *Dallas Tower Studies*

Reference-list number: 33

Reference: P.B. MacCready, T.B. Smith, and M.A. Wolf. Vertical Diffusion from a Low Altitude Line Source—Dallas Tower Studies, Vol. 1. Final Report MR161 FR-33 from Meteorology Research, 2420 North Lake Ave., Altadena, Calif. to U.S. Army Chemical Corps, Dugway Proving Ground. Contract DA-42-007-CML-504. Dec. 1961.

Only two title pages and pp. I-iii and 1-3 were available.

Principal object:
To relate measured diffusion characteristics of the cloud to observed turbulence, wind velocity, and temperature observations in a quantitative manner under a variety of meteorologic and release height conditions.

Site selection:
Shortly after the close of the Windsoc field observations, an Air Force-sponsored program of meteorologic observations was commenced on a television tower near Dallas and in the grid area previously used in the Windsoc studies. The availability of the instrumented television tower at Dallas provided a unique opportunity for the current studies.

Number and nomenclature of releases:
There were 37 releases in three test periods in April, June, and August 1961.

Test conditions:
The site was contained within a 30-mile radius of the 1,420-ft television tower at Cedar Hill, Tex., about 15 miles southwest of Dallas. Releases were from a light aircraft making cross-wind traverses between 1 and 7 miles upwind of the tower, over line lengths from 9 to 26 miles, at

heights between 360 and 1,050 ft above the tower base. Samplers were located on the tower and on five downwind lines (in directions south, southwest, west, northwest, and north from the tower) and one crosswind line (perpendicular to the northward sampling line and a distance of approximately 25 miles from the tower).

Test material:
No further data available.

Place of release:
No further data available.

Dates, times, and quantities of release:
No further data available.

Communities affected:
No further data available.

Distances from releases to affected communities:
No further data available.

Weather conditions:
No further data available.

Other materials released:
No further data available.

Maximum time-integrated concentrations:
No further data available.

Maximum time-integrated concentrations in any populated area:
No further data available.

Maximum concentrations:
No further data available.

APPENDIX B 251

Maximum concentrations in any populated area:
No further data available.

Other comments:
Only two title pages, three summary pages, and three pages of the document are available. All the information required should be in the full document.

DESCRIPTION OF TEST

Name of test: *Behavior of Aerosol Clouds within Cities*

Reference-list number: 35 (35A–F)

Reference: Behavior of Aerosol Clouds within Cities. Joint quarterly reports submitted by Stanford University and the Ralph M. Parsons Company to the U.S. Army Chemical Corps. Contracts DA-18-064-CML-1856 (Stanford) and DA-18-064-CML-2282 (Parsons).
    A. JQR No. 1, July-Sept. 1952
    B. JQR No. 2, Oct.-Dec. 1952
    C. JQR No. 3, Jan.-Mar. 1953
    D. JQR No. 4, Apr.-June 1953 (missing pp. 97, 98)
    E. JQR No. 5, July-Sept. 1953 (missing pp. 1-6, 32, 37)
    F. JQR No. 6, Oct.-Dec. 1953, 2 vols. (Vol. 2 missing pp. 144-166, 174)

Principal object:
To provide the field data required to estimate munitions requirements for the strategic use of chemical and biologic agents against cities.

Site selection:
The requirement was for climatologic, topographic, demographic, and physical similarities with Russian target cities with populations greater

than 100,000. The upper Mississippi Valley and adjacent areas presented the best match for climate and topography. The cities given consideration within this area were Oklahoma City, Kansas City, Omaha, Cincinnati, St. Louis, Chicago, Minneapolis, and Winnipeg. Population and physical characteristics of the cities suggested St. Louis, Winnipeg, and Minneapolis as the best choices, in that order. Tests were carried out in all three cities.

Within the cities, approximately 1 mile square test sites were selected on the basis of physical and meteorologic characteristics to represent types of areas within the city.

In Minneapolis:

Able Area: Residential-commercial. Approximately 1.5 miles from the central business district, bounded by 25th St., 35th St., 1st Ave. South, and Chicago Ave.. This area is now bisected by Route 65 (not present on contemporary maps).

Baker Area: River. Bisected by the Mississippi River, 2 miles south of Dog Area, roughly bounded by 6th Ave. Southeast to the northwest, 8th St. to the northeast, Walnut St. to the east, and 7th St. to the south.

Charlie Area: Open. Relatively flat area, 8 miles south of Minneapolis, bounded by Highway 5 and 100 on the south, by 34th Ave. South and Fort Snelling National Cemetery on the east, by 70th St. on the north, and by 24th Ave. South on the west.

Dog Area: City. Downtown, bounded by 3rd Ave. and the outlying railroad complex on the northwest, by the Mississippi River on the northeast, by Park Ave. on the southeast, and by 11th St. on the southwest.

Easy Area: Undeveloped. Approximately 10.4 miles north-northeast of Dog area, 6.4 miles north of city limits. Substituted for Charlie area, which was not available during summer. Bisected by Highway 8 running north-south, and by Ramsey County Road J running east-west.

In addition, there was at least one release during training in New Brighton approximately 8 miles from the city center (no further details are available).

In St. Louis:

How Area: Residential-commercial. Near the center of the St. Louis metropolitan section, approximately 2 miles west of the Mississippi River and approximately 1 mile from the center of downtown St. Louis. Bounded generally by Grand Blvd. on the west, Montgomery St. on the north, 22nd on the east, and Pine Blvd. on the south. For certain releases, these boundaries were somewhat extended.

Item Area: City. Downtown portion of St. Louis. Bounded by the Mississippi River on the east, Biddle St. on the north, 18th St. on the west, and Spruce St. on the south. For certain releases, these boundaries were somewhat extended.

These areas are distinct from those principally affected by the releases discussed in Refs. 17 and 43.

In Winnipeg:

Tare Area: City. The downtown commercial district and peripheral residential areas.

Uncle Area: Undeveloped. A 1-mile square area approximately 18 miles north of Winnipeg and 2 miles east of the north-south highway leading from Winnipeg to the Northern Provinces. The area was bounded by dirt roads, and was located 2 miles east of Stony Mountain, an 80-ft rise in otherwise flat land heavily timbered with scrub growth.

Number and nomenclature of releases:

In the Minneapolis winter tests, there were 63 releases during 23 tests. Tests were labeled FT0001 through FT0023 (FT0001 and FT0002 were training tests, with no releases), with releases indicated by appended letters (a, b, c . . .). There were two to four releases per test.

The Minneapolis summer program comprised 15 tests with 39 releases. Field tests were numbered FT0024 through FT0038, with releases indicated by appended letters (a, b, c . . .). There were two or three releases per test.

In the St. Louis winter tests, there were 17 field tests, during which tracer was released 35 times. Field tests were numbered FT1001 through FT1017 (FT1001 was a training test, with no releases), with releases indicated by appended letters (a, b, c . . .). There were one to three releases per test.

In the Winnipeg summer tests, there were 14 tests involving 36 releases. Field tests were numbered FT2001 through FT2014, with releases indicated by appended letters (a, b, c . . .). There were two or three releases per test.

Test conditions:
Releases were mostly from point sources (on the back of a truck, on a roof, or occasionally on the ground), occasionally from line sources (on the back of a moving truck), generally for approximately 5 min. Some releases were dual point sources, consisting of two point sources approximately 15 yd apart. Others were simultaneously from two point sources 100 to 400 yd apart. Sampling was on an approximate grid within a subsection of the selected area, modified by streets and conditions on the ground, using membrane filters. Each test generally used samplers at the same locations, but the release location and the sampler locations were changed from test to test. Citywide exposures were from extended line sources, again from the back of a truck carrying the FP disseminator. In the undeveloped (and unpopulated) areas Easy, Charlie, and Uncle, samplers were set along parallel lines downwind of the source. The nearest line of samplers was approximately 300 ft from the source.

Test material:
Ultraviolet FPs, New Jersey Zinc Company No. 2266 ZnCdS particles. There is no discussion within the available documents on the particle-size distribution, particle count, or dissemination efficiency. In JQR 5 (p. 73)

and JQR 6 (p. 69), during calculations of sedimentation losses, it is stated that the effective release particle density is 3×10^{10} particles per gram, and this value is used here.

Place of release:
Point sources were located at different locations within the test area for each test, and the sampling grid was changed for each test. Line sources were usually along one side of the test area (so that the sampling grid would be affected). In Minneapolis, the citywide line-source dispersion was from a truck-mounted generator running along State Highway 100 from CR 3 to CR 9 (FT0022), and from Broadway St. Northeast along 18th Ave. Southeast, dodging round the railway yards to Oak St. Southeast, along East River, across the river, and then along 31st Ave. South to East 49th St. (FT0023).

Dates, times, and quantities of release:
In the following tables, the abbreviations used are

- pt   point source
- line   line source (length of line given in the Separation or Length column)
- dual pt   two point sources, separated by about 15 yd
- two pt   separated point sources (separation given in the Separation or Length column).

The areas (in the Area column) are described under "Site selection" above. Missing entries indicate a lack of data (due to missing reports or failure of the reports to give the information), or nonrelevance. (In the last column, there should be an entry only for two-point or line sources.)

Minneapolis

Test	Date	Day	Time, CST	Area	Source Type	Mass, g	Dispersal Time, min	Separation or Length, yd
Training FT0001a	Unknown 01/15/53	Thu	Equip. check	New Brighton		Unknown		
FT0002a	01/16/53	Fri	Training	Able		0		
FT0002b	01/16/53	Fri	Training	Able		0		
FT0003a	01/19/53	Mon	20:32	Able	pt	7.7	5	
FT0003b	01/19/53	Mon	21:35	Able	pt	6.9	5	
FT0004a	01/21/53	Wed	20:10	Able	pt	5.8	5	
FT0004b	01/21/53	Wed	21:23	Able	pt	6.7	5	
FT0004c	01/21/53	Wed	22:43	Able	pt	9.1	5	
FT0005a	01/26/53	Mon	20:22	Baker	pt	10.4	5	
FT0005b	01/26/53	Mon	21:38	Baker	pt	8.9	5	
FT0006a	01/28/53	Wed	20:15	Baker	pt	9	5	
FT0006b	01/28/53	Wed	21:48	Baker	pt	9.6	5	
FT0007a	01/30/53	Fri	20:12	Baker	pt	8.3	5	
FT0007b	01/30/53	Fri	21:37	Baker	pt	8.7	5	
FT0007c	01/30/53	Fri	23:05	Baker	pt	8.3	5	
FT0008a	02/03/53	Tue	20:06	Able	pt	8.1		
FT0008b	02/03/53	Tue	21:34	Able	pt	9		
FT0008c	02/03/53	Tue	23:04	Able	pt	7.4		
FT0009a	02/09/53	Mon	20:17	Able	pt	12.2	5	
FT0009b	02/09/53	Mon	21:35	Able	pt	12.3	5	

(Continued)

Test	Date	Day	Time, CST	Area	Source Type	Mass, g	Dispersal Time, min	Separation or Length, yd
FT0009c	02/09/53	Mon	23:05	Able	pt	12	5	
FT0010a	02/12/53	Thu	00:20	Able	pt	13.3	5	
FT0010b	02/12/53	Thu	01:40	Able	pt	12.1	5	
FT0010c	02/12/53	Thu	03:05	Able	pt	11.5	5	
FT0010d	02/12/53	Thu	04:25	Able	pt	12.2	5	
FT0011a	02/15/53	Sun	14:05	Able	pt	11.8	5	
FT0011b	02/15/53	Sun	15:35	Able	pt	11.6	5	
FT0011c	02/15/53	Sun	17:10	Able	pt	10.9	5	
FT0012a	02/16/53	Mon	22:00	Charlie	pt			
FT0012b	02/16/53	Mon	22:54	Charlie	pt	9.7	5	
FT0012c	02/16/53	Mon	23:45	Charlie	pt			
FT0012d	02/17/53	Tue	00:50	Charlie	pt			
FT0013a	02/18/53	Wed	20:30	Charlie	pt			
FT0013b	02/18/53	Wed	21:35	Charlie	pt			
FT0013c	02/18/53	Wed	22:30	Charlie	pt			
FT0013d	02/18/53	Wed	22:55	Charlie	pt			
FT0014a	03/03/53	Tue	20:05	Able	pt	8.9	5	
FT0014b	03/03/53	Tue	21:25	Able	line	14.3	4.5	670
FT0014c	03/03/53	Tue	22:45	Able	line	18.2	5.17	670
FT0015a	02/24/53	Tue	20:06	Dog	pt	11	5	
FT0015b	02/24/53	Tue	21:55	Dog	pt	9.8	5	
FT0015c	02/24/53	Tue	22:35	Dog	pt	10.6	5	
FT0016a	02/27/53	Fri	20:27	Dog	pt			

258

(Continued)

Test	Date	Day	Time, CST	Area	Source Type	Source Mass, g	Dispersal Time, min	Separation or Length, yd
FT0009c	02/09/53	Mon	23:05	Able	pt	12	5	
FT0016b	02/27/53	Fri	21:25	Dog	pt			
FT0016c	02/27/53	Fri	22:35	Dog	pt	11.6	5	
FT0017a	03/07/53	Sat	00:53	Dog	pt	9	5	
FT0017b	03/07/53	Sat	02:11	Dog	pt	11.3	5	
FT0017c	03/07/53	Sat	03:40	Dog	pt	10.4	5	
FT0018a	02/23/53	Mon	20:15	Charlie	pt	11.7	5	
FT0018b	02/23/53	Mon	21:10	Charlie	pt			
FT0018c	02/23/53	Mon	22:12	Charlie	pt			
FT0018d	02/23/53	Mon	23:08	Charlie	pt	8	5	
FT0019a	03/04/53	Wed	20:05	Baker	line	23.5	5.17	500
FT0019b	03/04/53	Wed	21:25	Baker	line	19.3	5.5	500
FT0019c	03/04/53	Wed	23:05	Baker	line	21.3	6	
FT0020a	03/18/53	Wed	20:05	Able	line	68.5	8.23	3000
FT0020b	03/18/53	Wed	21:25	Able	pt	9.7	5	
FT0020c	03/18/53	Wed	22:45	Able	line	163	8.83	3000
FT0021a	03/21/53	Sat	00:45	Baker	pt	8.2	5	
FT0021b	03/21/53	Sat	02:15	Baker	pt	12.8	5	
FT0021c	03/21/53	Sat	03:50	Baker	pt	8.9	5	
FT0021d	03/21/53	Sat	05:15	Baker	pt	1.5	5	
FT0022a	03/24/53	Tue	19:59	Citywide	line	404	26.67	11440
FT0022b	03/24/53	Tue	23:00	Citywide	line	822.6	30	11440
FT0023a	04/28/53	Tue	20:00	Citywide	line	693.3	28.17	12320

(Continued)

Test	Date	Day	Time, CST	Area	Source Type	Mass, g	Dispersal Time, min	Separation or Length, yd
FT0009c	02/09/53	Mon	23:05	Able	pt	12	5	
FT0023b	04/28/53	Tue	22:05	Citywide	line	384	30.5	12320
FT0024a	08/21/53	Fri	21:40	Able	pt	18.1	5	
FT0024b	08/21/53	Fri	23:06	Able	pt	36.1	5	
FT0025a	08/23/53	Sun	13:09	Able	dual pt	81.8	5	
FT0025b	08/23/53	Sun	14:22	Able	dual pt	65.9	5	
FT0025c	08/23/53	Sun	15:35	Able	dual pt	62.3	5	
FT0026a	08/24/53	Mon	20:57	Able	dual pt	64.7	5	
FT0026b	08/24/53	Mon	22:15	Able	dual pt	71	5	
FT0026c	08/24/53	Mon	23:37	Able	two pt	47.9	5	225
FT0027a	08/25/53	Tue	21:00	Able	line	351	10.55	4840
FT0027b	08/25/53	Tue	22:48	Able	pt	20.1	5	
FT0027c	08/26/53	Wed	00:00	Able	line	340.3	10.68	4840
FT0028a	08/26/53	Wed	21:10	Dog	pt	25.4	5	
FT0028b	08/26/53	Wed	23:30	Dog	pt	21.9	5	
FT0028c	08/26/53	Wed	23:50	Dog	pt	21.5	5	
FT0029a	08/27/53	Thu	21:05	Dog	2 pt	56.3	5	233
FT0029b	08/27/53	Thu	22:20	Dog	dual pt	41.9	5	
FT0029c	08/27/53	Thu	23:35	Dog	two pt	42.4	5	230
FT0030a	08/28/53	Fri	21:05	Dog	line	141	4.85	2110
FT0030b	08/28/53	Fri	22:35	Dog	pt	25.8	5	
FT0030c	08/28/53	Fri	23:50	Dog	line	147.1	4.7	2110
FT0031a	08/30/53	Sun	13:20	Dog	dual pt	47.9	5	

260

(Continued)

Test	Date	Day	Time, CST	Area	Source Type	Mass, g	Dispersal Time, min	Separation or Length, yd
FT0009c	02/09/53	Mon	23:05	Able	pt	12	5	
FT0031b	08/30/53	Sun	14:37	Dog	pt	44.5	5	
FT0031c	08/30/53	Sun	15:45	Dog	pt	41.1	5	
FT0032a	09/10/53	Thu	21:15	Citywide	pt	150.8	5	
FT0032b	09/10/53	Thu	23:45	Citywide	line	807.1	24.8	
FT0033a	09/09/53	Wed	22:20	Easy	pt	30.7	5	
FT0033b	09/09/53	Wed	22:43	Easy	pt	24.5	5	
FT0033c	09/09/53	Wed	23:23	Easy	pt	38.2	5	
FT0034a	09/10/53	Thu	00:03	Easy	pt	18.5	5	
FT0034b	09/10/53	Thu	00:46	Easy	2 pt	63.7	5	
FT0034c	09/10/53	Thu	01:22	Easy	pt	43.7	5	
FT0035a	10/11/53	Sun	21:05	Citywide	pt	303.9	10	
FT0035b	10/11/53	Sun	23:20	Citywide	pt	330.4	10	
FT0036a	10/17/53	Sat	20:35	Citywide	pt	335.5	10	
FT0036b	10/17/53	Sat	23:17	Citywide	pt	157.4	5	
FT0037a	09/16/53	Wed	20:43	Able	pt	35.5	5	
FT0037b	09/16/53	Wed	22:05	Able	pt	18.5	5	
FT0038a	10/18/53	Sun	20:47	Citywide	pt	296.4	10	
FT0038b	10/18/53	Sun	23:05	Citywide	pt	305.3	10	
Total release						7920 g[a]		

[a]This total includes an allowance for missing data.

St. Louis

Test	Date	Day	Time, CST	Area	Source Type	Mass, g	Dispersal Time, min	Separation or Length, yd
FT0009c	02/09/53	Mon	23:05	Able	pt	12	5	
FT1001a	05/18/53	Mon	Training	How				
FT1001b	05/18/53	Mon	Training	How				
FT1002a	05/20/53	Wed	20:11	How	pt	12.6	5	
FT1002b	05/20/53	Wed	21:40	How	pt	10.1	5	
FT1003a	05/22/53	Fri	20:17	How	pt	9.4	5	
FT1003b	05/22/53	Fri	22:05	How	pt	11	5	
FT1004a	05/25/53	Mon	19:56	How	pt			
FT1004b	05/25/53	Mon	21:10	How	pt	9.6	5	
FT1004c	05/25/53	Mon	22:35	How	pt	10.1	5	
FT1005a	05/28/53	Thu	20:06	How	pt	12	5	
FT1005b	05/28/53	Thu	21:17	How	pt	6	3.66	
FT1005c	05/28/53	Thu	22:35	How	pt	12.9	5	
FT1006a	05/30/53	Sat	23:35	How	line	290.2	12.12	5100
FT1006b	05/31/53	Sun	01:26	How	pt	12.2	5	
FT1006c	05/31/53	Sun	03:35	How	line	95	11.2	5300
FT1007a	06/25/53	Thu	20:56	Citywide	line	760.2	23.12	10600
FT1007b	06/26/53	Fri	22:51	Citywide	line	835.3	25.35	10750
FT1008a	06/07/53	Sun	13:09	How	pt	12.3	5	
FT1008b	06/07/53	Sun	14:37	How	pt	12.6	5	
FT1008c	06/07/53	Sun	15:55	How	pt	11.5	5	
FT1009a	06/09/53	Tue	20:36	Item	pt	5	5	
FT1009b	06/09/53	Tue	22:19	Item	pt	9.6	5	

262

(Continued)

Test	Date	Day	Time, CST	Area	Source Type	Mass, g	Dispersal Time, min	Separation or Length, yd
FT1010a	06/11/53	Thu	21:01	Item	pt	8.25	5	
FT1010b	06/11/53	Thu	22:43	Item	pt	11.1	5	
FT1011a	06/13/53	Sat	23:50	Item	pt	9.3	5	
FT1012a	06/15/53	Mon	20:45	Item	pt	13.5	5	
FT1012b	06/15/53	Mon	22:27	Item	line	396.3	11.9	5600
FT1013a	06/18/53	Thu	21:15	Item	two pt	21.8	5	370
FT1013b	06/18/53	Thu	22:47	Item	dual pt	18.7	5	
FT1014a	06/20/53	Sat	23:40	How	two pt	19.5	5	345
FT1014b	06/20/53	Sat	01:00	How	dual pt	20	5	
FT1015a	06/20/53	Sat	03:00	Item	two pt	18.5	5	
FT1015b	06/20/53	Sat	04:10	Item	dual pt	25	5	
FT1016a	06/21/53	Sun	04:10	Item	two pt	22.1	5	233
FT1016b	06/21/53	Sun	15:40	Item	dual pt	22.2	5	
FT1017a	06/23/53	Tue	20:17	How	two pt	21.4	5	367
FT1017b	06/23/53	Tue	21:46	How	dual pt	22.9	5	
Total release						2775 g[a]		

[a]This total includes an allowance for missing data.

Winnipeg

Test	Date	Day	Time, CST	Area	Source Type	Mass, g	Dispersal Time, min	Separation or Length, yd
FT2001a	07/09/53	Thu	21:10	Tare	pt	24	5	
FT2001b	07/09/53	Thu	22:27	Tare	pt	13.3	5	
FT2001c	07/09/53	Thu	23:35	Tare	pt	14.9	5	
FT2002a	07/11/53	Sat	00:23	Tare	pt	13.4	5	
FT2002b	07/11/53	Sat	01:47	Tare	pt	16.7	5	
FT2002c	07/11/53	Sat	02:35	Tare	two pt	35.5	5	220
FT2003a	07/12/53	Sun	12:20	Tare	pt	18	5	
FT2003b	07/12/53	Sun	13:20	Tare	pt	10.1	5	
FT2003c	07/12/53	Sun	14:35	Tare	two pt	27.5	5	115
FT2004a	07/16/53	Thu	22:43	Tare	two pt	40.9	5	385
FT2004b	Missed sampling array							
FT2005a	07/21/53	Tue	21:05	Tare	line	369.5	9.1	3000
FT2005b	07/21/53	Tue	22:10	Tare	two pt	40.1	5	230
FT2005c	07/21/53	Tue	23:05	Tare	line	194.9	10.5	3000
FT2006a	07/25/53	Sat	13:05	Tare	line	308.5	6.7	2800
FT2006b	07/25/53	Sat	14:46	Tare	two pt	40.3	5	330
FT2006c	07/25/53	Sat	15:50	Tare	line	364.1	7.8	2800
FT2007a	Missed sampling array							
FT2007b	Missed sampling array							
FT2008a	07/14/53	Tue	22:45	Tare	pt	22.1	4.5	
FT2008b	07/14/53	Tue	23:50	Tare	pt	21.1	5	
FT2008c	07/15/53	Wed	01:05	Tare	two pt	29	5	220
FT2009a	07/23/53	Thu	20:35	Citywide	line	610.4	14.4	5800
FT2009b	07/23/53	Thu	23:07	Citywide	line	491	15.4	5800

263

264

(Continued)

Test	Date	Day	Time, CST	Area	Source Type	Mass, g	Dispersal Time, min	Separation or Length, yd
FT2010a	07/30/53	Thu	22:35	Tare	line	321.7	7.8	2800
FT2010b	07/31/53	Fri	01:25	Tare	two pt	19.5	5	280
FT2010c	Missed sampling array							
FT2011a	08/03/53	Mon	20:57	Uncle	two pt	39.8	5	150
FT2011b	08/03/53	Mon	21:45	Uncle	line	786	12.5	5456
FT2012a	08/03/53	Mon	22:47	Uncle	two pt	51.2	5	160
FT2012b	08/03/53	Mon	23:49	Uncle	pt	23.3	5	
FT2013a	08/02/53	Sun	14:35	Uncle	two pt	25	5	165
FT2013b	08/02/53	Sun	16:20	Uncle	line	530	13	5632
FT2013c	08/02/53	Sun	17:20	Uncle	two pt	21.5	5	185
FT2014a	08/01/53	Sat	12:35	Citywide	line	630.1	21.9	8800
FT2014b	08/01/53	Sat	14:53	Citywide	line	600.3	22.5	8800
Total release						5874 g[a]		

[a]This total includes an allowance for missing data.

Appendix B 265

Communities affected:
The releases took place primarily within the cities of Minneapolis, St. Louis, and Winnipeg. Most releases were within small sections of these cities, and affected relatively small areas, with plumes extending a few city blocks only. Some releases were "citywide," from the backs of trucks driven over distances of 3.3 to 6.5 miles along city streets or highways.

Distances from releases to affected communities:
Releases took place within the communities mentioned for most of the tests. Charlie and Easy areas near Minneapolis and Uncle area near Winnipeg appear to have been unpopulated and a considerable distance from any community.

Weather conditions:
Most Minneapolis winter releases were with snow-covered ground. Most Minneapolis summer releases were in clear or scattered cloud conditions, and no summer releases were with overcast or with other than dry ground conditions.

Other materials released:
No other releases are mentioned.

Maximum concentrations and time-integrated concentrations:
There were data reported for most of the dispersion tests, although some were never reported in the joint quarterly reports.

Maximum calculated concentrations and maximum measured time-integrated concentrations (exposures) for point and line sources depend very strongly on the distance from the source to the sampling point. The concentrations and exposures given in the following table vary substantially because of the varying distances to the nearest sampler. As mentioned above, the point-source releases gave rise to plumes that were generally detectable only for a few city blocks (less than 0.5 mile in most cases), with exponential reductions in concentration with distance. In each Minneapolis summer release, for example, the area within the 3.3 μg-min/m^3

contour was generally estimated to be in the range of 2×10^4 yd^2 to 3×10^5 yd^2 for the point-source releases and 5×10^6 yd^2 to 7×10^6 yd^2 for the citywide releases. The area within the 0.33 μg-min/m^3 contour was generally estimated to be 4 to 10 times these values. (The sampling network was not large enough to obtain more accurate estimates.) The areas affected by the (maximum) exposures reported in the following table were very small, probably only of the order of 100-1,000 yd^2.

Each test was carried out with a different pattern of samplers, and the release point was varied from release to release, so that it would be very difficult to accumulate the total dose at any individual point. However, for the different releases within each test, the samplers were (for the most part) at constant locations, so that exposures could be accumulated for the releases within tests. The changing locations and the small size of the point-source plumes ensured that the maximum (point source) exposures were not additive across tests (because only very small areas are affected by such maximum exposures). Additional line-source tests principally affected the areas close to the line sources, but potentially could affect the whole test area. The tests listed as citywide were generally larger line-source tests. Again, they principally affected areas immediately adjacent to and downwind of the source.

The approach taken here was designed to obtain a conservative estimate (overestimate) of the highest cumulative exposures. The highest measured exposure in each release was located and recorded, and the values for releases within each numbered test of a given type (point, line, citywide) were added, even if they were measured at different sample points, to obtain an estimate of the highest cumulative exposure in that test at any point. For the few cases in which different source types were used for separate releases in the same numbered test, the different source types were cumulated separately. Different point-source tests were assumed to affect different areas at the highest exposures, so the highest cumulative exposure from point-source tests within each named test area was selected to represent a conservative cumulative exposure estimate for such point-source tests within that test area. Line-source tests associated with a given named test area were assumed to affect that whole named test

APPENDIX B   267

area, so that cumulative exposures from line-source tests were added to the highest cumulative exposure from any point-source test within each test area. Finally, citywide tests were assumed to affect all test areas within each city (excluding unpopulated areas located many miles from the city).

The result of these calculations are estimates of maximum cumulative exposures within each named test area. The highest of these for populated and unpopulated test areas are listed in the following tables in the rows labeled "maximum populated" and "maximum unpopulated." (There were no tests in St. Louis in unpopulated areas.)

Concentrations were estimated for point sources by using the maximum measured exposure and dividing by the release period. The same approach was taken for the line and citywide sources, because some time-resolved data indicated that the plume was present for approximately the release period even in these cases, although it is likely that points close to the release lines experienced shorter immersion in higher concentration plumes (the line source consisted of a release at a constant rate from the back of a moving vehicle). The highest concentration estimate for any release was selected as an estimate of the maximum concentration ever experienced, because all releases occurred at different times. These estimates were made separately for the unpopulated and populated test areas and are reported in the following tables in the rows labeled "maximum populated" and "maximum unpopulated."

In the following table, a blank entry indicates lack of data (not reported in or missing from available copies of the reports). The fourth and fifth columns of the table are the cumulative exposures for a particular test; the fourth column (labeled "point") is for point-source tests, and the fifth column (labeled "area") is for line-source or citywide tests. The entries are placed next to the last release in each numbered test of the given source type. (For the few cases in which different source types were used in the same numbered tests, there will be an entry in both the fourth and fifth columns.). An allowance has been made for missing data within numbered tests, wherever such an allowance is relevant and possible. Missing exposures have been assumed to be equal to those for measured releases of the same source type within each numbered test, because releases of the same source type within each numbered test were very similar. Where the plume missed the sampling array, no estimate has been made, because no exposure could cumulate at any points on the sampling array.

The precision of the entries in the table is far higher than the accuracy. The accuracy for the Minneapolis winter data is, at best. a factor of 2. The St. Louis data, the Winnipeg data, and most of the Minneapolis summer data have, at best, an accuracy of a factor of 10, because most of the exposure data have been estimated from plotted but barely readable exposure isopleths.

Minneapolis

Test	Highest Exp., μg-min/m^3	Highest Conc., μg/m^3	Cumulative Highest Exp. (Point) μg-min/m^3	Cumulative Highest Exp. (Area) μg-min/m^3
FT0001a	0	0		
FT0002a	0	0		
FT0002b	0	0		
FT0003a	333	67		
FT0003b	1067	213	1400	
FT0004a	13	3		
FT0004b	29	6		
FT0004c	77	15	118	
FT0005a	261	52		
FT0005b	39	8	300	
FT0006a	65	13		
FT0006b	1257	251	1322	
FT0007a	407	81		
FT0007b	165	33		
FT0007c	55	11	627	
FT0008a				
FT0008b				
FT0008c				
FT0009a	16	3		
FT0009b	15	3		
FT0009c	27	5	58	

(Continued)

Test	Highest Exp., μg-min/m³	Highest Conc., μg/m³	Cumulative Highest Exp. (Point) μg-min/m³	Cumulative Highest Exp. (Area) μg-min/m³
FT0010a	93	19		
FT0010b	89	18		
FT0010c	39	8		
FT0010d	29	6	250	
FT0011a	28	6		
FT0011b	52	10		
FT0011c	85	17	164	
FT0012a				
FT0012b	1900	380		
FT0012c				
FT0012d			7600	
FT0013a				
FT0013b				
FT0013c				
FT0013d				
FT0014a	917	183	917	
FT0014b	88	20		
FT0014c	199	38		287
FT0015a	16	3		
FT0015b	104	21		
FT0015c	101	20	221	
FT0016a				
FT0016b				
FT0016c	17	3	50	
FT0017a	27	5		
FT0017b	103	21		
FT0017c	11	2	141	
FT0018a	3013	603		
FT0018b				
FT0018c				
FT0018d	5433	1087	16893	
FT0019a	99	19		
FT0019b	106	19		
FT0019c				308
FT0020a	86	10		
FT0020b	81	16	81	
FT0020c	200	23		286

(Continued)

Test	Highest Exp., μg-min/m³	Highest Conc., μg/m³	Cumulative Highest Exp. (Point) μg-min/m³	Cumulative Highest Exp. (Area) μg-min/m³
FT0021a	7	1		
FT0021b	6	1		
FT0021c	34	7		
FT0021d	6	1	53	
FT0022a	77	3		
FT0022b	208	7		285
FT0023a	6	0		
FT0023b	5	0		10
FT0024a	137	27		
FT0024b	167	33	303	
FT0025a	67	13		
FT0025b	50	10		
FT0025c	130	26	247	
FT0026a	100	20		
FT0026b	233	47		
FT0026c	83	17	417	
FT0027a	133	13		
FT0027b	67	13	67	
FT0027c	213	20		347
FT0028a	62	12		
FT0028b	60	12		
FT0028c	21	4	143	
FT0029a	142	28		
FT0029b	1520	304		
FT0029c	280	56	1942	
FT0030a	16	3		
FT0030b	59	12	59	
FT0030c	25	5		41
FT0031a	17	3		
FT0031b	40	8		
FT0031c	25	5	82	
FT0032a				
FT0032b				
FT0033a	250	50		
FT0033b	250	50		
FT0033c	250	50	750	
FT0034a				

APPENDIX B

(Continued)

Test	Highest Exp., μg-min/m³	Highest Conc., μg/m³	Cumulative Highest Exp. (Point) μg-min/m³	Cumulative Highest Exp. (Area) μg-min/m³
FT0034b				
FT0034c				
FT0035a	333	33		
FT0035b	333	33	667	
FT0036a				
FT0036b	17	3	33	
FT0037a	100	20		
FT0037b	67	13	167	
FT0038a	167	17		
FT0038b	250	25	417	
Maximum unpopulated		1087	17188	
Maximum populated		304	2615	

St. Louis

Test	Highest Exp., μg-min/m³	Highest Conc., μg/m³	Cumulative Highest Exp. (Point) μg-min/m³	Cumulative Highest Exp. (Area) μg-min/m³
FT1002a	50	10		
FT1002b	22	4	72	
FT1003a	83	17		
FT1003b	317	63	400	
FT1004a				
FT1004b	67	13		
FT1004c	50	10	175	
FT1005a	230	46		
FT1005b				
FT1005c	67	13	100	
FT1006a	25	2		
FT1006b	7	1	7	
FT1006c	13	1		38
FT1007a	12	1		
FT1007b				23
FT1008a	3	1		

(Continued)

Test	Highest Exp., μg-min/m³	Highest Conc., μg/m³	Cumulative Highest Exp. (Point) μg-min/m³	Cumulative Highest Exp. (Area) μg-min/m³
FT1008b	3	1		
FT1008c	3	1	9	
FT1009a				
FT1009b	17	3	34	
FT1010a	3	1		
FT1010b			6	
FT1011a	8	2	8	
FT1012a	13	3	13	
FT1012b	13	1		13
FT1013a	25	5		
FT1013b	83	17	108	
FT1014a	240	48		
FT1014b	167	33	407	
FT1015a	211	42		
FT1015b	1687	337	1898	
FT1016a	13	3		
FT1016b	13	3	27	
FT1017a	400	80		
FT1017b	667	133	1067	
Maximum populated		337	1935	

Winnipeg

Test	Highest Exp., μg-min/m³	Highest Conc., μg/m³	Cumulative Highest Exp. (Point) μg-min/m³	Cumulative Highest Exp. (Area) μg-min/m³
FT2001a	667	133		
FT2001b	67	13		
FT2001c	67	13	800	
FT2002a	67	13		
FT2002b	23	5		
FT2002c	667	133	757	
FT2003a	167	33		
FT2003b	333	67		
FT2003c	117	23	617	

(Continued)

Test	Highest Exp., μg-min/m³	Highest Conc., μg/m³	Cumulative Highest Exp. (Point) μg-min/m³	Cumulative Highest Exp. (Area) μg-min/m³
FT2004a	5000	1000		
FT2004b	5000			
FT2005a	100	11		
FT2005b	1253	251	1253	
FT2005c	83	8		183
FT2006a	67	10		
FT2006b	27	5	27	
FT2006c	100	13		167
FT2007a				
FT2007b				
FT2008a	83	19		
FT2008b	83	17		
FT2008c	233	47	400	
FT2009a	67	5		
FT2009b	33	2		100
FT2010a	100	13		
FT2010b	100	20	100	
FT2010c				100
FT2011a	67	13	67	
FT2011b	117	9		117
FT2012a	50	10		
FT2012b	7	1	57	
FT2013a	67	13		
FT2013b	67	5		67
FT2013c	667	133	733	
FT2014a	17	1		
FT2014b				33
Maximum unpopulated		133	917	
Maximum populated		1000	5583	

Other comments:
The cooperation of city officials was sought at meetings in St. Louis on August 27, 1952 (Mr. Lawrence F. Wood represented Mayor Darst, who was ill) and in Minneapolis on August 28, 1952 (Mayor Eric G. Hoyer). In both cases, other public officials were present, and the permission of

the mayors was subsequently obtained. The public officials were apparently misinformed of the reasons for the tests—they were to be informed that "the work was to obtain data pertinent to smoke screening of cities from aerial observation." This cover story was also used in Winnipeg.

In Minneapolis, there was some local interest in the operations. (The operations included meteorologic surveys that are part of these release experiments but are not otherwise noted here.) Initially, some samplers were "molested by curious passers-by, and several were actually found missing from stations." (All were returned by local citizens or by police who had been given the samplers.) The Minneapolis Tribune carried an article on 20 Jan. 1953 about the "little gray boxes that just sit on street corners, ticking and purring" and reported that "[t]hey could be used to measure the concentration in the air of a fine, harmless powder blown over the city." The newspaper also reported the smoke-screen story. Other papers followed with similar stories. Sampling equipment was subsequently chained and locked to trees, lamp poles, and similar objects, and few molestations of consequence subsequently occurred. There was no difficulty obtaining part-time and full-time personnel to set out and collect samplers and technicians to read the sample filters.

In St. Louis, meetings were held with city officials again in April 1953 to outline the summer test program. Meetings were also held with officials of Monsanto Chemical Company, Socony-Vacuum Oil Company, Granite City Steel Corp., and the Board of Aldermen of Granite City, Ill. Full cooperation of the industrial firms was obtained, and permission was granted for use of company properties for field-test sites. Public interest in the field-test phase was much less than that in Minneapolis. Only a few small articles were printed in the local press. There was considerable public interest when Forest Park had to be used as a wiresonde site (for the meteorologic surveys), but no incidents of consequence were noted. There was difficulty getting personnel for the experiments in the tight labor market and with "the generally disinterested attitude," resulting in rapid turnover, a limitation of the scope of some experiments, and the necessity to discard some data because of poor quality.

In Winnipeg, civil defense and city officials and surrounding municipalities were cooperative, and considerable interest was shown by the press and public, particularly in the first wiresonde ascents when a teenager attempted to use the kytoon as a bow-and-arrow target. (Subsequent use of more remote locations eliminated difficulties.) In addition, "[t]wo articles in a Winnipeg paper debated the feasibility and advisability of using smoke screens as a means of civil defense. Another article reported the action of the Labor Progressive Party of Canada in writing a letter to the Mayor of Winnipeg in which objection was voiced to the conduct of a smoke-screen study in Winnipeg." Personnel were available, often through the civil-defense organization, and "a generally good attitude was exhibited," although the time of tests had to be moved to 15 min after sunset, rather than the preferred 1 h after sunset to get an adequate part-time force so late at night.

All weekday tests in all cities took place at night (after 6 p.m. and before 6 a.m.). Some weekend tests were performed during the day, usually the early afternoon.

In the winter tests in Minneapolis, analysis of samplers located inside and outside the Clinton School during 12 releases and in five houses in which penetration studies were conducted during seven releases gave the results shown in Table B-2. For office buildings in winter, the results shown in Table B-3 were obtained. Summer results for houses and offices are given in Table B-4 and Table B-5.

Similar penetration results for St. Louis and Winnipeg are shown in Table B-6 and Table B-7.

TABLE B-2 Penetration into Houses and the Clinton School (Percentage of Outside Doses) in Winter

		Relative Doses	
Houses	Number of Samples	Median (%)	Extreme range (%)
Basement	19	13	0-58
First Floor	13	11	0-200
Second Floor	10	2	0-41
GROSS	42	11.5	0-200
Clinton School			
Ground Floor	36	23.5	0-100
First Floor	13	27	11-43
Second Floor	22	22.5	0-100
GROSS	71	23	0-100

TABLE B-3 Penetration into Office Buildings in Minneapolis in Winter—Inside Doses Relative to Outside Doses at the Same Height

		Relative Doses		
Height Range (Floors)	Number of Samples	Median (%)	Interquartile Range (%)	Extreme Range (%)
Sub-basement[a]	27	13	0-31	0-173
Basement[a]	31	11	0-30	0-143
1-4	20	24	14-52	0-834
5-8	32	16	4-34	0-1400
9-12	27	10	0-25	0-652
13-17	10	38	12-76	0-124
GROSS	147	15	0-37	0-1400

[a]Basement and sub-basement doses were compared with ground-level outside doses.

TABLE B-4 Penetration into Houses (Percentage of Outside Doses), Minneapolis, Summer

Height Range (floor)	Number of Samples	Relative Doses		
		Median (%)	Interquartile Range (%)	Extreme Range (%)
Basement	22	52	29-100	0-662
First Floor	12	86	60-96	19-400
Second Floor	9	59	21-129	0-408
Third Floor	1	54	—	—
Gross	44	58	34-98	0-662

TABLE B-5 Penetration into Office Buildings in Minneapolis in Summer—Inside Doses Relative to Outside Doses at the Same Height

Height Range (ft)	Number of Samples	Relative Doses		
		Median (%)	Interquartile Range (%)	Extreme Range (%)
Basement[a]	76	30	10-57	0-150
6-50	18	22	7-71	0-130
51-100	21	36	22-50	0-91
101-200	10	52	31-226	0-540
GROSS	125	31	11-57	0-540

[a]Basement doses were compared with ground-level outside doses.

TABLE B-6 Penetration of Buildings (Windows Closed) in St. Louis (May and June)—Relative Dose at the Same Height.

Height Range (ft)	Number of Samples	Relative Doses		
		Median (%)	Interquartile Range (%)	Extreme Range (%)
Basement[a]	53	33	14-100	0-1800
6-50	26	42	16-100	1-1250
51-100	37	45	16-78	0-175
101-200	26	26	3-77	0-877
GROSS	142	34	13-93	0-1800

[a]Basement doses were compared with ground-level outside doses.

TABLE B-7 Penetration into Residences in Winnipeg (Inside Doses Relative to Outside Ground-level Doses)

Height Range (ft)	Number of Samples	Relative Doses		
		Median (%)	Interquartile Range (%)	Extreme Range (%)
Basement	11	82	52-96	0-212
First Floor	8	86	60-124	0-1400
Second Floor	8	64	28-88	0-125
Gross	27	80	52-98	0-1400

A technical summary of the data from these studies was to be published by Stanford, but it is not available for this report.

A final administrative and operational report was to be published by the Ralph M. Parsons Company, giving an enumeration of all field tests, but it is not available for this report.

No complete summary of all the tests performed in any city is available, and the results of some tests were not reported anywhere in the series of JQRs. Critical pages of some of the JQRs are missing, so that certain details (particularly the amounts of material released) are not available for some tests. There are occasional slight differences in details between the available summary tables and the detailed information in the appendices of the JQRs. The tables above report the most consistent set of values available.

Individual digits of some of the numbers might be incorrect, because of unreadable script in the available reports. Many of the values given are highly uncertain estimates based on the locations of plotted exposure contours, together with partial reading of some digits of values written next to individual sample points (where these were legible). It is also likely that there were transcription errors in the original reports, because they were compiled by hand.

Some required pages of the available documents were not reproduced in their entirety in the available copies (e.g., JQR 6, Vol. 1, pp. 25-26).

Reproduction of most of the appendices, and the fold-out pages in particular, in JQR 6 was inadequate. As mentioned above, exposures are primarily estimates based on contours visible on the fold-out pages for Winnipeg, St. Louis, and the summer series in Minneapolis.

DESCRIPTION OF TEST

Name of test: *Stanford* (first field test)

Reference-list number: 36 (BMRs 1-13 and final report)

Reference: S.W. Grinnell, W.A. Perkins, F.X. Webster. BMR 11 submitted by Stanford University to the U.S. Army Chemical Warfare Service Research and Development Program. Contract W-18-035-CWS-1256, Sept.-Oct. 1947.

BMRs. 1 through 13 and the final report are all available. Only BMR 11 describes any field release. The others imply small (milligram quantity) releases into an air chamber, and "field work of a minor nature."

Principal object:
This was the first of a contemplated series, with general objectives: to demonstrate the feasibility of use of the fluorescent method, and the capability of detecting a few 1-5 μm particles per liter; to demonstrate the suitability of available equipment; to correlate meteorologic measurements with observations of aerosol travel; to determine the lateral diffusion rate of aerosol clouds; to determine optimum distribution of sampling stations; and to demonstrate long-range travel of aerosol clouds.

Site selection:
Speed was of the essence in demonstrating the basic technique on a scale of a few hundred meters. A 500 × 300 open field conveniently located to the immediate rear of the project buildings (prefab steel barracks located just to the rear of the main chemistry buildings at Stanford, 1946).

Number and nomenclature of releases:
A single release. Equipment was all experimental and not ready for repeated trials.

Test conditions:
The newly designed multi-jet continuous feed dispenser was used to release the material from a point source over 165 s. Samplers of various types were set up at 70 m (four samplers) and 170 m (six samplers). All that was needed was a steady wind with a westerly component.

Test material:
ZnCdS, New Jersey Zinc Company, No. 2266. Primary particle size range from 1 to 8 μm diameter, with number mean diameter approximately 2 μm. No calibrations had been performed at the time of this experiment; a particle release rate of 1×10^{10} ppg are assumed on the basis of subsequent calibrations of similar (possibly identical) material.

Place of release:
In the field described under "Site selection."

Dates, times, and quantities of release:
Approximately 1700 on 15 Oct. 1947. Total release of 0.826 g.

Communities affected:
Unknown but could be determined. The direction of travel of the plume is known, but the location of the test field with respect to local populations is not described (but could be determined).

APPENDIX B

Distances from releases to affected communities:
Unknown but might be determinable (see "Communities affected"). Greater than approximately 500 m, the distance across the open test field.

Weather conditions:
Not described, except for wind conditions. We may presume that it was not raining.

Other materials released:
None described.

Maximum time-integrated concentrations and maximum concentration:
Only one sampler (at 174 m from the release) provided usable dosage data (although at least one other sampler showed the presence of the cloud in its vicinity).

The measured exposure was approximately 5 µg-min/m^3. With a release time of 165 s, this measurement corresponds to an average concentration of 2 µg/m^3—an upper bound on the exposure of any community.

Other comments:
These reports document the genesis of the FP tracer technique using ZnCdS. There is mention in Report No. 9 of receipt of 5 lb of ZnCdS and 5 lb of DuPont fluorescent zinc silicate. The ZnCdS was shown to be much the superior tracer in a subsequent report. We cannot presume that the 5 lb were all used, because the authors used this material under subsequent continuing contracts.

The contract number is given incorrectly in the reference list. The bimonthly reports were performed under W-18-035-CWS-1256.

DESCRIPTION OF TEST

Name of test: *San Francisco*

Reference-list number: 37 (37A–G)

Reference: Quarterly reports submitted by Stanford University to the U.S. Army Chemical Corps Research and Development Program. Contract DA-18-108-CML-450.
    A. QR 1, Feb.-Mar. 1950 (tests at Palo Alto)
    B. QR 2, May-June 1950 (tests at Dugway Proving Ground)
    C. QR 3, Aug.-Sept.-Oct. 1950 (San Francisco and adjacent area)
    D. QR 4, Missing
    E. QR 5, Feb.-Mar.-Apr. 1951 (measurement of urban temperature gradients)
    F. QR 6, May-June-July 1951 (meteorologic studies)
    G. QR 7, Aug.-Sept.-Oct. 1951 (meteorologic studies; FP size measurement)

Principal object:
FE 15-16. To repeat the observations of FE 14 that showed that the presence of an inversion allowed significant exposures at large distances (4.5 miles).

FE 17-25 To evaluate: the probable hazard from airborne clouds of pathogenic material released at Dugway; the travel of aerosol clouds at Dugway compared with that over populated areas; and the comparative behavior of clouds of biologic material and FPs.

FE 26-30 To verify results of previous tests in the San Francisco Bay area; to obtain quantitative information on aerosol travel over much larger populated areas, and under different conditions from earlier tests; and to obtain a comparison between an inert and a biologic aerosol along several miles of travel over populated areas.

Site selection:
No specific reasons are stated for selecting Palo Alto, Dugway, and San Francisco.

Number and nomenclature of releases:
QR 1: Mention is made of previous reports that are not available. Included are some details of FE 14.
QR 1: Two releases, designated FE 15 and FE 16.
QR 2: Nine releases, designated FE 17-25.
QR 3: Six releases in five experiments, designated FE 26-30; FE 29 had two consecutive releases.

Test conditions:
Palo Alto
FE 15-16 released from the back of a truck along lines to the northwest of Palo Alto. Seven drum samplers were operated over a 2-h period, giving 1-min time discrimination. In FE 16, a membrane filter was operated 6 yd from the release line.

Dugway
(Note: The locations of the following named features were not identified because of the lack of the accompanying explanatory charts. Modern public maps identify several features and road names that are named in the reports, so the names might still be current.)

During the following four daytime tests, the wind was from the northwest.
FE 17, 18, and 20: Releases were along the GPI-III road.
FE 21: Release was from a fixed point on the Lincoln Highway north-northwest of Target X.

During the following three daytime tests, the wind was from the south to southwest.
FE 22: Release was made "south of the main road starting 14.5 miles from camp."

FE 23: The same as that in FE 22, but moved to take account of a wind shift. The first 3.5 miles of the release was in the vicinity of Granite Peak.

FE 24: Release line not identifiable from available documents but should have been in a location similar to FEs 22-23.

During all the daytime tests, the aerosol cloud and the dust cloud raised by the vehicle emitting the aerosol cloud was observed to rise immediately in the unstable air. The closest samplers did not measure the highest exposures.

The following two tests were performed at night.

FE 19: Release on a 0.8-mile arc of an 800-yd circle around target X, starting at the east pill box. Wind direction was southeast to south-southeast.

FE 25: The same as that in FE 19, with a 1-mile release line. Wind direction was south-southeast to south.

Samplers were located at distances from several hundred yards to 14 miles downwind, with some samplers moved into position during or after the release on the basis of wind measurements.

San Francisco

Releases were from a 100-ft Mine Layer using three or four generators mounted 15 ft above the water, and traveling north to south along a 2- to 6-mile line approximately parallel to the shore and 2 to 10 miles from the shore. For FE 29, two release lines were used, starting from 9.5 miles west of San Francisco. In every case, the release lines were traversed in approximately 30 min.

There were 43 sampling stations, 10 of which were equipped with drum samplers giving 1-min resolution. Most remaining stations were equipped with filter samplers, although up to five stations per experiment had no FP measurement equipment (stations were also equipped with samplers for the biologic material released).

Appendix B

Test material:

New Jersey Zinc Company No. 2266 (ZnCdS). The emission rate stated in QR 1 (Palo Alto) corresponds to 1.82×10^{10} ppg. In QR 2 (Dugway), the particle count is given as 3×10^{10} ppg, although this count appears to be based on an estimated mass mean diameter of about 2.5 µm. In QR 3 (San Francisco), the particle count is said to be 6×10^{10} ppg, a count based on a mass mean diameter of 2 µm.

Place of release:

Palo Alto
FE 15 and FE 16. Along lines northwest of Palo Alto.

Dugway
See "Test conditions" above.

San Francisco
See "Test conditions" above. The average distance offshore of the releases were 2.5 miles for all experiments except the two releases of FE 29, which were 10 miles out.

Dates, times, and quantities of release:

FE	Date	Start Time	Period, min	Length, miles	Amount, g
Palo Alto					
15	03/10/50	15:30	16	3.58	500.1
16	03/14/50	15:25	14	3.43	475.5
Total					975.6
Dugway					
17	07/01/50	14:40	17	3.5	880.5
18	07/11/50	10:00	17	3.6	778
19	07/19/50	03:59	8	0.8	372.7
20	07/29/50	14:18	19	3.5	917
21	07/30/50	15:23	25	pt	625
22	08/03/50	12:00	20	5.5	1012.4
23	08/03/50	15:08	55	10.4	1961.8

24	08/04/50	14:26	24	5.8	1086.2
25	08/04/50	23:48	11	1	610.6
Total					8244.2
San Francisco		**PDST**			
26	10/20/50	14:30	28	5.85	3426
27	10/22/50	14:00	30.1	7.09	4328
		PST			
28	10/25/50	17:00	28	2.7	5263
29	10/26/50	16:15	29	2	2677
	10/26/50	17:11	27.2	2.42	2546
30	10/27/50	17:00	30	2.88	4200
Total					22,440

Communities affected:
FE 15-16: Palo Alto.

FE 17-25: At Dugway test ground. Unclear what was affected. Possibly one or more of the local small communities.

FE 26-30: San Francisco and the Bay area.

Distances from releases to affected communities:
FE 15-16: Not clear from available documents. The experiment was designed so that the main cloud passed through the city of Palo Alto.

FE 17-25: It is noted for FE 17 that "a significant dosage was found at the village 14 miles from the source", and "This is the first time at Dugway that any airborne material has been sampled more than a mile or two from its point of release, and is even more noteworthy in view of the fact that the test was run during the day when conditions were least favorable for long distance travel."

FE 26-30: Aerosols were generated 2 to 10 miles west of San Francisco, on a course approximately parallel to the shore.

Weather conditions:

Palo Alto
FE 15 and FE 16: Inversions at 3600 ft and 2800 ft.

Dugway
FE 17, 18, 20, 21: Performed during the day under strong lapse conditions, with the wind from the northwest.

FE 22-24: Performed during the day under strong lapse conditions, with the wind from the south to southwest. During FE 23, sand and thunder storms were encountered along the release line.

FE 19, 25: Nighttime releases. During FE 19, the lower atmosphere was almost certainly neutral because of the high wind. During FE 25, there was almost certainly an inversion.

San Francisco
FE 26, 29, and 30 were in clear weather with the typical Bay-area onshore winds with inversion conditions at various heights over the ocean but sufficient heating over the city to provide lapse conditions at least to altitudes of several hundred to 1,000 ft or so. FE 27 had clear sky conditions, but a low, dense offshore fog, again with some inversion conditions over the ocean but lapse conditions over land. FE 28 had fog covering San Francisco to Oakland, keeping the inversion in place.

Other materials released:
FE 19 (at Dugway): BG was released simultaneously using clusters of E48R2 bombs dropped from the air.

FE 25 (at Dugway): SM was released from a cluster of E48 bombs fired statically.

FE 26-30: A biologic material was released simultaneously with the FP.

Maximum time-integrated concentrations and concentrations:
The following table gives estimates of maximum time-integrated concentrations (exposures) for each experiment, together with estimates of the

maximum short-term concentrations, where such estimates could be made. In addition, a cumulative estimate is made for the whole series of reported experiments.

A maximum exposure for experiment FE 14 is given because it is mentioned in QR 1, and no further documentation of the early FE series is available.

The maximum concentrations, where available, are from 1-min resolution drum samplers. For each experiment over San Francisco, the highest measured 1-min concentration has been increased in proportion to the total exposure (i.e., the highest measured 1-min concentration at any drum sampler has been multiplied by the ratio of highest total exposure measured anywhere to the total exposure measured by that drum sampler) because drum samplers were not placed at every sampling location. For San Francisco, the cumulative total exposure is the largest measured at any single location. (It is dominated by the single large value measured in FE 28 during a fog.) For Palo Alto and Dugway, samplers were at different locations for each experiment, so the sum of the maximum total exposures is an overestimate for the cumulative maximum exposure.

FE	Populated Area		Unpopulated Area	
	Max. Exp., μg-min/m^3	Max. Conc., μg/m^3	Max. Exp., μg-min/m^3	Max. Conc., μg/m^3
Palo Alto				
14	2.7			
15	2.1	0.5		
16	0.3	0.1		
Cumulative	2.4	0.5		
Dugway				
17	0.03 (?)		0.5	
18			3.1	
19			10.0	
20			0.1	
21			1.5	0.5

APPENDIX B   289

(Continued)

	Populated Area		Unpopulated Area	
	Max. Exp.,	Max. Conc.,	Max. Exp.,	Max. Conc.,
FE	μg-min/m³	μg/m³	μg-min/m³	μg/m³
22			0.2	
23			0.1	
24			0.0	
25			63.2	
Cumulative			<78.6	0.5 (?)
San Francisco				
26	17.3	1.0		
27	4.2	1.1		
28	410.3	15		
29	13.9	0.8		
30	18.9	0.7		
Cumulative	436	15		

Other comments:

The first of these QRs reported two tests, numbers 15 and 16, of a series. Evidently, numbers 1 through 14 are reported in other material that is not available.

The maps provided in QR 1 are readable but not decipherable. They might be decipherable if the reports for the previous contract were available.

There is a reference to tests subsequent to those reported in QR 2 by the Detrick test team. No details are given.

In QR 2, figures 9, 11, 12, 14-17, 20, 21, 25-27, 35, 36, and 39-41 are missing. Figure 9 is the site map.

In QR 2, Charts A through I from Appendix A, are missing. These show sampling locations, release locations, exposures, meteorologic measurement area, balloon trajectories, and other information.

In QR 3, Charts A through E giving details of locations, are missing. There is a notice stating that these charts are available for viewing at the Technical Information Center, Dugway Proving Ground, Utah (801)831-3564. Page 23 is missing.

DESCRIPTION OF TEST

Name of test: *Collection Efficiency of the Rotorod FP Sampler*

Reference-list number: 41

Reference: F.X. Webster. Collection Efficiency of the Rotorod FP Sampler. Technical Report TR 98. Aerosol Laboratory, Metronics Associates, 3201 Porter Drive, Palo Alto, Calif. Contract DA 42-007-CML-543 for the U.S. Army Chemical Corps Research and Development Program. 31 Jan. 1963.

Principal object:
To investigate and define the collection efficiency of the FP collector rod for the Rotorod sampler with the coating technique as employed at Dugway Proving Ground.

Site selection:
Apparently for convenience—the field-test sites were not exactly located but were usually not more than a few hundred feet from the Metronics building.

Number and nomenclature of releases:
Seven field experiments, with four releases each, for a total of 28 releases (most of the releases required two distinct passes; some required as many as four passes). The field experiments were labeled FE 134 through FE 140, and the trials within each field experiment were labeled A, B, C, and D.

Appendix B

Test conditions:
A line source of FP particles was made from a high-speed blower-type aerosol generator (5-8 g/min) mounted on the tail gate of a station wagon driven slowly along a route approximately 150 to 750 ft upwind of the nearest sampler. Two sampler stations were used, with an array of Rotorod samplers surrounded by reference membrane filter samplers. Two to four passes of the generator were used in each trial, two generally sufficing but as many as four being required in a few cases under highly variable wind conditions. (Test samplers were monitored for enough particle collection to give accurate statistics.)

Test material:
Various types of FP were used. Abbreviations are
  MMD mass mean diameter
  Manf NJZ = New Jersey Zinc Company; USRC = U.S. Radium Corp.
  Date Date of manufacture
  DPG Dugway Proving Ground.

Color	Type	MMD	ppg	Lot	Manuf.	Date	Trial
Yellow	2266	1.8	7.0×10^{10}	MA 104	NJZ	1954	FE 137
Yellow	2266	2.4	3.3×10^{10}	SAL 8B	NJZ	1952	FE 134 & 135
Green	3206	2.4	3.3×10^{10}	H324-2	USRC	1961	FE 138
Yellow	2267	3.1	1.6×10^{10}	1339-2	USRC	1962	FE 140
Yellow	2267	3.4	1.2×10^{10}	DPG 12-21	USRC	1960	FE 139
Yellow	2267	3.8	0.9×10^{10}	H310 A-E	USRC	1960	FE 136

Place of release:
A few hundred feet from the Metronics building at 3201 Porter Drive, Palo Alto, Calif.

Dates, times, and quantities of release:
The quantities released were not specified, nor were any data supplied that allow reasonably exact computation. The release rate (in grams per minute and grams per foot) was given for each field experiment, but the time or distance of the release continued for each trial was not given, nor was even the number of passes required for each trial. The following ta-

ble gives a rough estimate obtained by assuming the disseminator was on for one-half the total period listed for the experiments. The value so obtained appears to be a reasonable estimate when considering also the length of the line source likely to have been used (about 2 or 3 times the distance to the nearest sampler) and the speed of the disseminator.

FE	Date	Start Time, PST	End Time, PST	Estimated Amount, g
134	01/26/62	15:15	16:32	193
135	03/27/62	15:18	15:56	154
		PDT	PDT	
136	05/09/62	14:47	15:37	200
137	05/14/62	15:21	17:15	291
138	07/17/62	16:59	17:45	145
139	08/21/62	15:15	16:06	219
140	11/16/62	15:44	16:43	206
	Cumulative			1408

Communities affected:
Palo Alto, around the Metronics building.

Distances from releases to affected communities:
Releases were within the community.

Weather conditions:
Wind speeds generally in the range 8 to 14 mph, although a few trials were conducted with wind speeds as low as 3 mph, and a few others in winds as high as 18 mph.

Other materials released:
No other materials were mentioned.

Maximum time-integrated concentrations in any populated area:
These concentrations were measured at 150 to 750 ft from the line source. It is likely that the experiment was set up so that there were no buildings between source and samplers.

APPENDIX B

FE	Trial	Max. Exp., µg-min/m³	FE	Trial	Max. Exp., µg-min/m³
134	A	20	138	A	111
	B	94		B	56
	C	17		C	48
	D	37		D	80
135	A	237	139	A	86
	C	51		C	64
	D	75		D	75
136	A	41	140	A	90
	B	38		B	66
	C	28		C	93
	D	34		D	55
137	A	19			
	B	24		Cumulative	1676
	C	14			
	D	16			

The cumulative estimate has been obtained by summing the individual values for each experiment, because most of the experiments were apparently carried out at the same location.

Maximum concentrations in any populated area:
No time-resolved data were available for estimating concentrations.

Other comments:
This contract was to improve the efficiency estimate for the Rotorod as used at Dugway. None of the Dugway studies included among the references used Rotorod samplers.

Reference is made to CML-543 Final Report (presumably from Metronics) giving tabulations of ppg and MMD for various FP, particularly the "12-series lots" used by Dugway. That might be Ref. 48.

FP Lot H324-2 is stated to have been used in Dugway Trial 502 B-7, but that is not in the references.

Reference is made to Stanford Aerosol Laboratory QR 111-10, July-Aug.-Sept. 1956, Contract DA-42-007-403-CML-111, although only in the context of theoretic analyses.

All the tracers listed above "have been used as tracers in atmospheric diffusion studies." H324-2 and DPG 12-21 have been "used extensively at Dugway."

Appendix C

Correspondence from the Army

DEPARTMENT OF THE ARMY
OFFICE OF THE DEPUTY CHIEF OF STAFF FOR OPERATIONS AND PLANS
WASHINGTON, DC 20310-0400

REPLY TO
ATTENTION OF

DAMO-FDB 6 May 1996

MEMORANDUM FOR Dr. Kulbir Bakshi, National Research Council/National Academy of Sciences, 2101 Constitution Avenue, N.W., Washington, DC 20418

SUBJECT: Zinc Cadmium Sulfide Dispersion Testing Documents

1. The Army has reviewed all known relevent zinc cadmium sulfide documents for declassification. Documents have been declassified in accordance with Army Regulation 380-86, Classification of Chemical Warfare and Chemical and Biological Defense Information.

2. Documents remaining classified, IAW AR 380-86, para 1.c.(12) Employment concepts, are available for review by individuals with the appropriate clearance and a need to know.

3. All of the relevant declassified documents have been provided to Dr. Kulbir Bakshi, program director of the National Research Council Committee on Toxicology. The Committee on Toxicology has identified some missing information in their review (enclosure 1). This enclosure will be forward to Aberdeen Proving Ground, Dugway Proving Ground, and the United States Army Chemical School, in a further attempt to locate the documents.

4. POC this action is CPT William M. Barnett, (703) 697-7001.

CARMEN J. SPENCER
COL, GS
Chief, Chemical and NBC
Defense Division

APPENDIX C

DEPARTMENT OF THE ARMY
OFFICE OF THE DEPUTY CHIEF OF STAFF FOR OPERATIONS AND PLANS
400 ARMY PENTAGON
WASHINGTON DC 20310-0400

REPLY TO
ATTENTION OF

DAMO-FDB   26 July 1995

MEMORANDUM FOR Dr. Kulbir Bakshi, (HA 354), National Academy of Sciences, 2101 Constitution Avenue, N.W., Washington, D.C. 20416

SUBJECT: Review of Declassified Large Area Coverage (LAC) Document

1. The Army has reviewed all relevent zinc cadmium sulfide documents for declassification. All known documents have been declassified in accordance with Army Regulation 380-86, Classification of Chemical Warfare and Chemical and Biological Defense Information.

2. Documents remaining classified, IAW AR 380-86, para 1.c.(12) Employment concepts, are available for review by individuals with the appropriate clearance and a need to know.

3. POC this action is CPT William Barnett, (703) 697-7001.

JOHN C. DOESBURG
COL, GS
Chief, Chemical and NBC
Defense Division

Appendix D

Interaction of Zinc and Cadmium and Toxicity of Zinc Cadmium Sulfide Activators

Interaction of Zinc and Cadmium and Toxicity of Zinc Cadmium Sulfide Activators

INTERACTION OF ZINC AND CADMIUM

ZINC IS AN ESSENTIAL nutrient and is toxic only at very high doses. Therefore, the subcommittee focused its review of the potential toxicity of ZnCdS on its most toxic component, cadmium, and examined the effect of zinc only in regard to its potential interactions with cadmium.

There are only two studies in the literature that investigated the toxic interaction of zinc and cadmium by the inhalation route. In rats exposed to cadmium chloride by inhalation, simultaneous exposure to zinc oxide prevents fatalities (Oldiges and Glaser, 1986) and lung cancer (Oldiges et al., 1989). The subcommittee believes that the results of these studies are not relevant to zinc cadmium sulfide exposures for two reasons. First, ZnCdS is a sintered compound. The sintered compound reportedly does not contain free zinc sulfide or cadmium sulfide, because the sintering process is highly efficient. Second, in the unlikely event that ZnCdS breaks down into its components, the exposure would be to such small amounts of the components, a biological response is unlikely.

TOXICITY OF COPPER AND SILVER

Copper or silver was added to ZnCdS as an activator at concentrations of less than 100 parts per million (ppm).

Copper is an essential nutrient that is incorporated into numerous enzymes; the National Research Council's recommended dietary allowance (RDA) for copper is 2-3 mg/d (NRC 1980). Because the ZnCdS formulation contained copper at less than 100 ppm (Sheila Fabiano, USR Optronix, personal commun., 1995), doses of copper much lower than the RDA would have been associated with exposure to ZnCdS. Therefore, the subcommittee concluded that copper toxicity from the Army's tests should not be a concern.

Silver is only toxic at high levels. The Occupational Safety and Health Administration permissible exposure level (PEL) for silver is 0.01 mg/m^3 (OSHA 1989). Because the ZnCdS formulation contained silver at less than 100 ppm (Sheila Fabiano, USR Optronix, personal commun., 1995), the subcommittee concluded that doses of silver associated with exposure to ZnCdS would have been too low to warrant concern.

REFERENCES

NRC (National Research Council). 1980. Recommended Dietary Allowances. 9th Rev. Ed. Washington, D.C.: National Academy Press. 185 pp.

Oldiges, H., and U. Glaser. 1986. The inhalative toxicity of different cadmium compounds in rats. Trace Elem. Med. 3:72-75.

Oldiges, H., D. Hochrainer, and U. Glaser. 1989. Long-term inhalation study with Wistar rats and four cadmium compounds. Toxicol. Environ. Chem. 19:217-222.

OSHA (Occupational Safety and Health Administration). 1989. Air Contaminants. Final Rule 29 CFR. Fed. Regist. 54(12):2332-2983.

Appendix E

Public Meeting Agendas

NATIONAL RESEARCH COUNCIL
BOARD ON ENVIRONMENTAL STUDIES
AND TOXICOLOGY
2101 Constitution Avenue   Washington, D.C. 20418

COMMITTEE ON TOXICOLOGY

TEL: (202) 334-2897
FAX: (202) 334-1393

PUBLIC MEETING

SUBCOMMITTEE ON ZINC-CADMIUM SULFIDE

Committee on Toxicology
Board on Environmental Studies and Toxicology
Commission on Life Sciences
National Research Council
National Academy of Sciences

Ramada Inn Bayfront
Austin/Dallas Room
601 North Water Street
Corpus Christi, Texas
October 18, 1995

DRAFT AGENDA

10:00 am - 10:30 am   Welcome, description of the National Research Council study on zinc-cadmium sulfide, and briefing on explanation of administrative procedures for the conduct of the public meeting
Rogene F. Henderson, Ph.D., Subcommittee Chair

10:30 am - 12:30 pm   **Presentations and Questions/Answers**

John H. Winchester
Corpus Christi, Tex

C. Lee Skaggs
Corpus Christi, Tex.

Pattie Floerke
Corpus Christi, Tex.

The National Research Council is the principal operating agency of the National Academy of Sciences and the National Academy of Engineering to serve government and other organizations. The Board on Environmental Studies and Toxicology is responsible to the National Research Council through the Commission on Life Sciences and the Commission on Geosciences, Environment, and Resources.

Father Patrick Donohoe
Immaculate Conception Church
Taft, Tex.

Linda Valencia
Portland, Tex.

Ginger Pope, daugher
and
Everett Karel
Corpus Christi, Tex.

Alice Pullin
Taft, Tex.

Hector Valle
Taft, Tex.

Philip Sensinger
Corpus Christi, Tex.

John Villanueva
Corpus Christi, Tex.

Reverend Marian Fletcher
First United Methodist Church
Taft, Tex.

12:30 pm   ***ADJOURN***

NATIONAL RESEARCH COUNCIL
BOARD ON ENVIRONMENTALK STUDIES
AND TOXICOLOGY
2101 Constitution Avenue   Washington, D.C. 20418

COMMITTEE ON TOXICOLOGY

TEL: (202) 334-2897
FAX: (202) 334-1393

PUBLIC MEETING

SUBCOMMITTEE ON ZINC-CADMIUM SULFIDE

Committee on Toxicology
Board on Environmental Studies and Toxicology
Commission on Life Sciences
National Research Council
National Academy of Sciences

Grand Wayne Center
Exhibit Hall East
120 West Jefferson Boulevard
Fort Wayne, Indiana 46802
July 31, 1995

DRAFT AGENDA

10:00 am - 10:30 am   Welcome, description of the National Research Council study on zinc-cadmium sulfide, and briefing on explanation of administrative procedures for the conduct of the public meeting
Rogene F. Henderson, Ph.D., Subcommittee Chair

10:30 am - 12:30 pm   Session I

The Honorable Paul Helmke
Mayor of Fort Wayne, Ind.

Dave Herbst
Fort Wayne, Ind.

Thomas Childers
Fort Wayne, Ind.

Mary Beth Pettyjohn
Fort Wayne, Ind.

Judy Anderson
Woodburn, Ind.

The National Research Council is the principal operating agency of the National Academy of Sciences and the National Academy of Engineering to serve government and other organizations. The Board on Environmental Studies and Toxicology is responsible to the National Research Council through the Commission on Life Sciences and the Commission on Geosciences, Environment, and Resources.

APPENDIX E

Harold G. Baker
Fort Wayne, Ind.

Sheila Urbine
Fort Wayne, Ind.

Barbara L. Johnson
Fort Wayne, Ind.

Mr. and Mrs. Tony Kassing
Fort Wayne, Ind.

Joseph Gibbons, M.D.
Fort Wayne, Ind.

Charlene Bloom
Fort Wayne, Ind.

John T. Federspiel
Fort Wayne, Ind.

Daniel L. Federspiel
and his son, Fred
Fort Wayne, Ind.

Tamra Fraser
Pendleton, Ind.

Lorene Sheckles
Pleasant Lake, Ind.

12:30 pm - 1:30 pm   *LUNCH*

1:30 pm - 3:00 pm   Session II

Ruth Arnold
Fort Wayne, Ind.

Carl E. Wallis
Fort Wayne, Ind.

Marilyn Swinehart
Fort Wayne, Ind.

Raymond Parker
Fort Wayne, Ind.

Carol Pritchard
Fort Wayne, Ind.

Deborah Snyder
Akron, Ind.

Sonia Myers
Fort Wayne, Ind.

Marie Martin
Ossian, Ind.

Gloriadell Uptgraft
Fort Wayne, Ind.

Norma Jean Steiss
Huntertown, Ind.

Michael Zeis
Fort Wayne, Ind.

Karla White
Fort Wayne, Ind.

William Bloch
Fort Wayne, Ind.

Kim Mills
(city unknown)

Bea Ebbinghaus
Noblesville, Ind.

Wylann K. Taylor
and her husband
Fort Wayne, Ind.

3:00 pm - 4:00 pm	Session III
	Questions and Answers
4:00 pm	***ADJOURN***

NOTE: Total time allowed for each presentation is 5-7 minutes (depending on number of speakers). Sequencing of speakers was based on the order in which requests were received. Written material will also be accepted.

APPENDIX E

NATIONAL RESEARCH COUNCIL
BOARD ON ENVIRONMENTALK STUDIES
AND TOXICOLOGY
2101 Constitution Avenue   Washington, D.C. 20418

COMMITTEE ON TOXICOLOGY

SUBCOMMITTEE ON ZINC-CADMIUM SULFIDE

TEL: (202) 334-2897
FAX: (202) 334-1393

Committee on Toxicology
Board on Environmental Studies and Toxicology
Commission on Life Sciences
National Research Council
National Academy of Sciences

Park Inn International
Conference Room: North and Center Forum
1313 Nicollet Mall
Minneapolis, Minnesota
May 25, 1995

PUBLIC MEETING

PRELIMINARY AGENDA

10:00 am - 10:20 am   Welcome, Description of the National Research Council Study on Zinc-Cadmium Sulfide, and Administrative Procedures for the Workshop
Rogene F. Henderson, Ph.D., Subcommittee Chair

10:20 am - 12:30 pm   Session I

Ms. Diane Gorney
President of Children of the 50s
Minneapolis, MN

Mr. John B. Heimkes
Bloomington, MN

Dr. Ian A. Greaves
Head, Environmental and Occupational Health
University of Minnesota, Minneapolis, MN 55455
"U.S. Army Risk Assessment and Health Concerns of Minneapolis Residents"

Mrs. Darleen E. Stein
Bloomington, MN

Ms. Barbara Sitz
Minneapolis, MN

The National Research Council is the principal operating agency of the National Academy of Sciences and the National Academy of Engineering to serve government and other organizations. The Board on Environmental Studies and Toxicology is responsible to the National Research Council through the Commission on Life Sciences and the Commission on Geosciences, Environment, and Resources.

Mr. Richard Meixner
Bloomington, MN

Ms. Sandy Swanson
Center City, MN
Ms. Terry Haugan
Minneapolis, MN

Dr. Leonard Cole
Rutgers University
Newark, NJ
"The Zinc-Cadmium Sulfide Tests: Risk, Ethics, and Response"

Ms. Kathy Daniels
Lakeville, MN

Mr. Harry F. Hubel
St. Paul, MN

Ms. Teresa M. Haugan
Richfield, MN

Mr. James Sanders Hurd
Little Canada, MN

Mrs. Thomas A. Hamilton (Gwen)
Lexington, MN

12:30 pm - 1:30 pm   *LUNCH*

1:30 pm - 3:00 pm   Session II

Ms. Carol Thomas Haken
Director
Children's Environmental Institute
Forest Lake, MN

Mr. John D. Nagel
Minneapolis, MN

Mr. Michael E. Moen
Director
Division of Disease Prevention and Control
Minnesota Department of Health
Minneapolis, MN
"The Role of Minnesota Department of Health in the Zinc-Cadmium Sulfide Issue"

APPENDIX E 309

                    Mr. Bruce Watson
                    Roseville, MN

                    Mr. William T. Rose
                    Mahtomedi, MN

                    Ms. Cindy J. Wall
                    Minneapolis, MN

                    Ms. Sharon Sandness
                    Minneapolis, MN

                    Ms. Bev-Nii Anderson
                    Environmental Specialist for Tribal Council
                      of Leech Lake Reservation
                    Cass Lake, MN

                    Mr. Jamie Eidsmoe
                    Cass Lake, MN

                    Mr. Donald D. McNeil
                    Shakopee, MN

3:00 pm - 4:00 pm   <u>Session III</u>

                    Open Discussion

4:00 pm   *ADJOURN*

NOTE: Total time allowed for each presentation is 5-8 minutes (depending on number of speakers). Written material will also be accepted.

Appendix F

Sampling and Analytic Methods for Zinc Cadmium Sulfide

Sampling and Analytic Methods for Zinc Cadmium Sulfide

THE CONCENTRATIONS OF airborne fluorescent particles of ZnCdS in the Army atmospheric-dispersion studies were measured with impingement and filtration methods. Two of those methods are thoroughly described by Leighton and others (1965), but the methods used at specific locations are not described in detail in Army risk-assessment documents.

FILTER SAMPLING

Leighton and others (1965) describes the use of membrane filters for collection of ZnCdS samples in atmospheric-dispersion experiments. The cellulose acetate-nitrate membrane filters are mounted in an open-faced holder with a cowl. The filter is dyed to provide a dark background for counting particles under ultraviolet illumination, and most of the collected particles are deposited on the smooth upper surface of the filter. Both 25-mm- and 50-mm-diameter filters are used; they have deposition areas of 2 and 13 cm^2, respectively. The flow rate is typically about 5 L/min per square centimeter of deposition area. Collection efficiency of the filters is virtually 100%. (Collection efficiency is the number of particles deposited on the collection medium from an air sample divided by the number of particles in an equal volume of ambient air at the sample location averaged over the same time that the sample is taken.)

ROTOROD SAMPLING

The Rotorod consists of 2 thin metal rods coated with silicone grease that are attached to the shaft of a small electric motor by a cross arm (Leighton and others 1965). As the rods move through the air, particles touch their surfaces and are retained by the silicone grease. The sampler described by Leighton and others (1965) had a pair of 0.38 × 60-mm collecting surfaces, a rotation radius of 60 mm, and a rotation speed of 2,400 rpm.

The sampling rate of the Rotorod sampler is the volume of air swept out by the sampling rods per unit time. Thus, it is equal to the rod cross-sectional area times its tangential speed—41 L/min for the sampler described by Leighton and others. Correction of Rotorod particle counts for collection efficiency is essential because the collection efficiency of this sampler is generally lower than that of filter sampling and is a strong function of particle size. The collection efficiency can be determined by sampling ZnCdS aerosols with both a Rotorod sampler and a sampler with known collection efficiency. The Rotorod collection efficiency is the uncorrected concentration measured with the Rotorod divided by the efficiency-corrected concentration measured by the other method. For several lots of ZnCdS 2266, collection efficiency ranged from 28% for a lot with a mass median diameter (MMD) of 1.8 μm and 7.9×10^{10} particles per gram to 73% for a lot with an MMD of 3.1 μm and 1.6×10^{10} particles per gram (Leighton and others 1965).

TAPE SAMPLING

Gelman paper tape samplers were used for some of the Army experiments. The air sample is drawn through a paper filter tape. Periodically, the tape advances, collecting a series of sequential particle samples. Particles in each sampling "spot" on the tape are counted.

PARTICLE COUNTING

Collection media are illuminated with ultraviolet light, usually from a mercury-arc lamp, providing high-intensity light at about 366-nm wavelength. For counting, a microscope with $100 \times$ magnification is usually used. Counting fields are established with an eyepiece reticule, and the number of particles is counted separately for each field.

The size of the smallest particle discernible with this method is a function of the illumination, the background characteristics, the visual acuity of the counter, and the microscope configuration. A typical lower limit of particle diameter detectable by the ultraviolet-illumination and fluorescent-counting method described by Leighton and others (1965) is 0.5 μm. Thus, no particles smaller than that size were counted in these experiments. But particles smaller than 0.5 μm make up only a small fraction of the ZnCdS mass, and on the basis of the particle size distribution of the ZnCdS used at Corpus Christi (Smith and Wolf 1963), this should not create more than a 1% negative bias in the estimation of particle concentration.

REFERENCES

Leighton, P.A., W.A. Perkins, S.W. Grinnell, and F.X. Webster. 1965. The fluorescent particle atmospheric tracer. J. Appl. Meteorol. 4:334-348.

Smith, T.B., and M.A. Wolf. 1963. Vertical Diffusion from an Elevated Line Source over a Variety of Terrains. Part A. Final Report. Contract DA-42-007-CML-545. Prepared by Meteorology Research, 2420 North

Lake Ave., Altadena, Calif., for the U.S. Army Dugway Proving Ground, Dugway, Utah.

Appendix G

Review of AEHA Risk-Assessment Reports on Zinc Cadmium Sulfide

Review of AEHA Risk-Assessment Reports on Zinc Cadmium Sulfide

THE ARMY'S HEALTH risk assessment for Minneapolis, MN, Corpus Christi, TX, Fort Wayne, IN, and St. Louis, MO, addresses the potential adverse health effects resulting from the Army's atmospheric-dispersion tests with ZnCdS. Its conclusion was that because of the brief exposure period and concentrations used, the dispersion tests should not have posed any adverse health effect for people in the test areas. The Army recognizes that there was a large degree of uncertainty associated with its estimates of human health risk, as in any risk assessment, and that the calculated estimates should be taken not as "absolute estimates of risk but rather as conditional estimates."

The subcommittee's evaluation of the Army Environmental Hygiene Agency (AEHA) reports on risk assessment of ZnCdS is as follows.

APPROPRIATENESS OF USING CADMIUM TOXICITY IN RISK-ASSESSMENT REPORTS

In the absence of toxicity data on ZnCdS, AEHA used the toxicity data on cadmium in a worst-case scenario. For an accurate and defensible assessment of potential health hazards resulting from an exposure to a chemical substance, it is important that particular characteristics of the substance be known. The strength and defensibility of the health assessment is greatly enhanced by the inclusion of information on the physical and chemical properties of the substance, its toxicokinetics and bioavailability, the type of toxic response that it elicits, the exposure concentration and duration necessary to produce the response, and the population at risk. Often, such data are lacking and the toxicologist must use other available toxicity data to assess the health risk accurately. An appropriate surrogate may be used—one of the individual components of the substance in question or a closely related substance with known toxicity. Use of such surrogates is especially appropriate if the physical and chemical properties of the two agents are similar, if the toxicity and toxicokinetics of the surrogate would be expected to mimic closely those of the test agent, and if judgment indicates that this approach would likely overestimate rather than underestimate the risk.

The National Research Council Committee on Toxicology (NRC 1988) and the US Environmental Protection Agency (EPA) have both used such an approach successfully. For example, chromium and chromium compounds have been used as surrogates for lithium chromate because little information was available on the toxicity of lithium chromate and because chromium is its most-toxic component, and benzene has been used as a surrogate for gasoline because benzene is its most-toxic component and EPA wanted to use a worst-case scenario.

The situation is similar for ZnCdS. There are few studies on the general toxicity, bioavailability, and toxicokinetics of ZnCdS, so it is necessary to consider the use of existing toxicity databases such as those on cadmium sulfide, zinc sulfide, and cadmium and zinc and their salts. Because zinc is an essential element for humans and animals and is generally considered to be relatively nontoxic in small amounts, the subcommittee concluded that the Army's approach of estimating the toxic potential of ZnCdS by using toxicity data on cadmium compounds as a surrogate was appropriate and could be considered to be conservative and prudent and to constitute a worst-case scenario for estimating risk associated with exposure to ZnCdS.

DISPERSION METHODS

CORPUS CHRISTI

ZnCdS was disseminated from an airplane flying parallel to the coast off shore of Port Aransas (AEHA 1994a). Nine single-release tests were performed over a 6-d period. Observed-concentration data on 2 of these tests are missing. ZnCdS was released at the rate of 1.5 lb/mile over a 30-mile course, producing a line source with a strength of 1.11×10^9 particles/ft (Smith and Wolf 1963). The tracer was released at an altitude of 500 ft in all experiments except for tests 15 and 17, whose release altitudes were 750 and 1,000 ft. Rotorod samplers were placed at 1-mile intervals along a 25-mile-long line downwind of and perpendicular to the line source at its midpoint.

MINNEAPOLIS

ZnCdS tracer was released from both stationary and mobile sources (AEHA 1994b). The winter program consisted of 23 field tests with 63 releases. Data are available on only 20 tests, which comprised 52 releases. The summer program consisted of 15 field tests with 39 releases, and data are available on all these tests. The winter program began on January 19, 1953, and lasted until April 28, 1953. Tests were performed in 4 areas of the city ("Able," "Baker," "Charlie," and "Dog") that were selected because of their varied topographic and land-use characteristics; ZnCdS was dispersed from a point source either on a stationary vehicle or on a rooftop. Two tests were also performed on a citywide basis; ZnCdS was dispersed from a moving vehicle (line source). The summer program ran from August 21, 1953 to September 18, 1953. Point-source and dual-point-source releases were made from rooftops in 3 areas of the city ("Able," "Dog," and "Easy"), and 4 citywide tests were made with ZnCdS dispersed from moving vehicles. Filter samplers were placed outside and in residences, schools, and commercial buildings. The sampling methods and networks are not described in detail in the Army risk assessment (AEHA 1994b).

FORT WAYNE

There were 23 field tests comprising 75 releases during the summer of 1964 and the winter of 1965-1966 (Hilst and Bowne 1966). There were 2-4 releases for each test; 2 airplanes were used. The planes released ZnCdS at 2.5 lb/mile along a 20-mile straight line perpendicular to the mean wind, upwind of the city. The first plane released yellow ZnCdS (ZnCdS2267, lot H-395), and the second released green ZnCdS (ZnCdS3206, lot H-396, and one release of lot H-391). After the initial release, the plane returned to the field for reloading. Later releases were made after the plumes of the first releases had passed through the sampling area. Cassette-filter samples were collected at 25 random locations in the Ft. Wayne sampling network. Each filter sampler was set up to take 10 30-min sequential samples and 1 full-period integrated sample. Some 250 Rotorod samplers were used to collect integrated samples along 5 lines across the city: 1 at the upwind edge of the city, 1 at the downwind edge, and 3 dividing the city according to major land-use categories. The 50 samplers in each line were spaced as evenly as possible. In addition, Gelman paper tape samplers, programmed to collect sequential 5-min-average samples, were placed at 10 locations to detect the times of arrival and departure of the tracer plume.

ST. LOUIS

Sixteen field tests comprising 35 releases were carried out from May 20, 1953, through June 25, 1953 (AEHA 1994c). Tests were performed in 2 areas representing different topographic and land-use characteristics: the "How" area near the center of the St. Louis metropolitan section about 2 miles west of the Mississippi River and 1 mile from downtown, and the "Item" area in the middle eastern portion of the metropolitan area. They included 19 single-point and 10 simultaneous-dual-point releases from stationary vehicles or rooftops. In addition, a citywide test used 2 line-source dispersals from moving vehicles. The Army risk assessment does not describe the sampling methods and sampling network in detail (AEHA 1994c). Samples were collected with the membrane filter method (J. Kirkpatrick, US Army Center for Health Promotion and Preventive Medicine, Personal commun., Feb. 8, 1996). One release did not yield useful data, because the samplers were not downwind of the release; this release was not considered in the Army risk assessment.

EXPOSURE ANALYSIS

Army risk assessments of ZnCdS were based on direct transport from source to human receptor. Inhalation was the only route of entry that the Army considered important. The general approach was to estimate the exposure in each city from air-sampling data. The risks associated with these exposures were then examined. The methods used to estimate exposures were somewhat different for each city.

CORPUS CHRISTI

Only the integrated particle-number concentration in particle-minutes per cubic meter from the Rotorod samplers was available for analysis (AEHA 1994a). The Rotorod samplers used in the Corpus Christi experiments had a calculated sampling rate of 41.8 L/min. The average sampling efficiency was found to be about 75%, compared with a filter-sampling method assumed to have an efficiency of 100% (Smith and Wolf 1963). An average effective sampling rate of 30.8 L/min was calculated on the basis of the sampling efficiencies. However, the same sampling rate as that used in earlier Dallas Tower experiments, 33 L/min, was used in the calculations for determining ZnCdS concentrations in Corpus Christi because sampling rates determined at Corpus Christi were variable and on the average were not distinguishably different from the Dallas sampling rate (Smith and Wolf 1963).

The highest ZnCdS concentration was observed at the sampling station closest to the source. In the AEHA (1994a) risk assessment, the integrated concentration from this station summed over all 9 tests, 232.7 particle-min/L, was used as a conservative estimate of exposure throughout the area. The particle-count exposure was converted to mass exposure by

APPENDIX G 321

using the number of particles per gram from Smith and Wolf (1963), 2.16×10^{10} particles/g. The average mass concentration was obtained by dividing the integrated concentration by the sampling time. The actual sampling time was not available, so the sampling time used in a similar study, 300 min, was used in the Army calculations. The computed ZnCdS concentration was 3.59×10^{-5} mg/m^3, or a cadmium concentration of 7.2×10^{-6} mg/m^3, assuming that ZnCdS is 20% cadmium by weight.

The Army risk assessment estimated an integrated cadmium exposure equivalent to 0.029 µg by inhalation of particles transported directly from source to human receptor. That value assumes a breathing rate of 0.8 m^3/h, 100% deposition efficiency in the lung (as a worst-case scenario; a more reasonable assumption would be that the deposition efficiency is 10-15%), and an exposure time of 300 min. The assumed sampling and exposure times cancel out and have no effect on the integrated-exposure estimate.

MINNEAPOLIS

Unlike the Corpus Christi study, the Minneapolis experiments varied in sources and receptor locations from test to test. The methods used to analyze concentration data for the Minneapolis experiments are described briefly in the Army risk assessment (AEHA 1994b). The approach taken was to plot isopleths of integrated concentration in particle-minutes per liter for each test on a Minneapolis map. The representative isopleth for the maximal concentration in each area and for each release was used to represent the dosage for that release. Isopleths used were from 100 to 100,000 particle-min/L with intervals for each power of 10 between (logarithmic intervals). For each test site, the areas of the regions between the 1,000- and 9,999-particle-min/L isopleths and between the 10,000- and 99,999-particle-min/L isopleths were reported (AEHA 1994b).

The Army then calculated the integrated exposure in milligram-minutes per cubic meter, the average concentration of ZnCdS in milligrams per cubic meter, and the average concentration of cadmium in milligrams per cubic meter corresponding to values of integrated count concentration at each isopleth, shown here in Table G-1. The conversions from particle-minutes per liter followed the procedure used in the Army risk assessment for Corpus Christi, TX, discussed above, assuming a sampling duration of 60 min.

TABLE G-1 Army-Calculated Concentrations for Each Isopleth Value of Integrated Particle-Count Concentration, Minneapolis, MN (AEHA 1994b)

Count Concentration (particle-min/L)	Sampling Time (min)	Integrated ZnCdS Concentration (mg-min/m^3)[a]	ZnCdS Concentration (mg/m^3)[a]	Cadmium Concentration (mg/m^3)[a]
100	60	4.63×10^{-3}	7.72×10^{-5}	1.54×10^{-5}
1,000	60	4.63×10^{-2}	7.72×10^{-4}	1.54×10^{-4}
10,000	60	4.63×10^{-1}	7.72×10^{-3}	1.54×10^{-3}
100,000[b]	60	4.63×10^{0}	7.72×10^{-2}	1.54×10^{-2}
247[c]	60	1.14×10^{-2}	1.91×10^{-4}	3.8×10^{-5}

[a]Values rounded from 4 or 5 significant figures to 3 significant figures.
[b]100,000 isopleth occurred only 1 time in only 1 area.
[c]Integrated count concentration reported at Clinton School.

The Army risk assessment states that the isopleths were the only exposure data, except for data from Clinton School, available from the "Minneapolis report," presumably the Chemical Corps reports referred to in the risk assessment (AEHA 1994b). The risk assessment states that the maximal integrated concentration obtained outside the Clinton School was 247 particle-min/L; this value was used to represent exposure at the Clinton School.

It is unlikely that the maximal reported concentrations in Minneapolis represent an upper bound for the outdoor exposures by direct inhalation, inasmuch as some people might have been closer to the source or more directly downwind of the source than any of the sampling stations.

Fort Wayne

The Army risk assessment for Fort Wayne gives estimates for 1-h, 8-h, and chronic (integrated) exposures (AEHA 1995). For the 1-h estimated exposure, the Rotorod sampler results were used because they gave the highest reported count concentrations. The highest reported integrated count, considering both yellow and green releases—yellow ZnCdS at 1,878 particle-min/L from the February 4, 1966, experiment—was selected for the 1-h exposure value in the Army risk assessment. The actual sampling times were not available, and the Army assumed a 38-min sampling time. That was the shortest period between releases, so it provides a conservative estimate of the individual sampling duration.

For the estimated 8-h exposures, the integrated concentrations for each release and for both colors of FP were summed to obtain the total daily maximal exposure. The maximal daily exposure over all the days was then selected as the 8-h exposure estimate: 2,035 particle-min/L from the December 5, 1965, experiment. The Army estimated the sampling time to be 150 min, but no rationale for this was presented.

To estimate chronic exposure, the Army plotted isopleths of integrated concentrations summed over the entire experimental program. The highest isopleth was 80 particle-min/L. The area on the high side of the 80-particle-min/L isopleths could be as high as 90 particle-min/L because the interval between isopleths is 10 particle-min/L. Thus, 90 particle-min/L was used as the chronic-exposure estimate.

The integrated concentration values in particle-minutes/liter for each of the 3 exposure times were converted to cadmium concentrations in milligrams per cubic meter by using the assumed sampling duration, a weight percentage of cadmium in ZnCdS of 20%, and the particles-per-gram values given in Hilst and Bowne (1966). The yellow ZnCdS had 1.32×10^{10} particles/g, and the green, 1.45×10^{10} particles/g. The mean of these, 1.385×10^{10} particles/g, was used for converting count to weight units when the sum of the yellow and green ZnCdS concentrations was used. The Army calculated the 1-h and chronic concentrations by dividing the integrated concentrations, shown in Table G-2, by the corresponding sampling times and particles-per-gram values. That approach yields a conservative value for the time-weighted average direct-transport inhalation exposures because it assumes that the concentration during the sampling period persisted beyond the sampling period. It is more likely that the direct-transport exposure by inhalation persists only during the time required to pass by a person downwind of the source; this is roughly the same as the sampling time. The average concentration in terms of weight of cadmium per volume of air was calculated by multiplying by 0.2, the Army's assumed weight fraction of cadmium in ZnCdS.

TABLE G-2 Army Estimates of Time-Averaged Concentrations for Fort Wayne, IN, Atmospheric Tracer Experiments (AEHA 1995)

Type of Exposure	Integrated ZnCdS Concentration (particle-min/L)	Sampling Time (min)	ZnCdS Concentration (mg/m^3)[a]	Cadmium Concentration (mg/m^3)[a]
1-h	1,878	38	3.75×10^{-3}	7.50×10^{-4}
8-h	2,035	150	1.47×10^{-4}	1.96×10^{-4}
Chronic	90	38	1.71×10^{-4}	3.42×10^{-5}

[a]Values rounded from 4 significant figures to 3 significant figures.

The 8-h average concentration reported in the Army risk assessment cannot be calculated with the approach applied to the 1-h and chronic concentration estimates. Also, the reported cadmium concentration is higher than the reported total FP concentration. It appears that there are errors in calculating the 8-h concentrations.

ST. LOUIS

In the Army risk assessment for St. Louis, air-monitoring data were analyzed with the same approach used for data from Minneapolis. The isopleths for St. Louis ranged from 10 to 10,000 particle-min/L (AEHA 1994c). The concentrations corresponding to these isopleth values are reproduced here in Table G-3.

TABLE G-3 Army-Calculated Concentrations for Each Isopleth Value of Integrated Particle-Count Concentration, St. Louis, MO (AEHA 1994c)

Count Concentration (particle-min/L)	Sampling Time (min)	Integrated ZnCdS Concentration (mg-min/m^3)[a]	ZnCdS Concentration (mg/m^3)[a]	Cadmium Concentration (mg/m^3)[a]
10	60	4.63×10^{-4}	7.72×10^{-6}	1.54×10^{-6}
100	60	4.63×10^{-3}	7.72×10^{-5}	1.54×10^{-5}
1,000	60	4.63×10^{-2}	7.72×10^{-4}	1.54×10^{-4}
10,000	60	4.63×10^{-1}	7.72×10^{-3}	1.54×10^{-3}

[a]Values rounded from 4 or 5 significant figures to 3 significant figures.

ADJUSTMENTS TO EXPOSURE ESTIMATES

It was noted that the conversion from ZnCdS to cadmium concentration is in error. This computation assumed that 20% of the mass of ZnCdS is cadmium, whereas actually only 15.6% is. Thus, the exposure estimates based on this concentration are 28% too high. Smith and Wolf (1963) discuss the effect of disseminator efficiency on the number of particles per gram. The efficiency was determined to be 39%, so the number concentration of particles per gram of dust dispersed produced by the disseminator is 39% of the number concentration per gram produced in the laboratory. A truly conservative approach to estimating exposure would be to decrease the number of particles per gram used in these calculations by 39%, increasing the weight concentration by a factor of 2.6. Taking into account the corrections for the weight fraction of cadmium in ZnCdS and for the disseminator efficiency, we recommend that the Army direct inhalation-exposure estimates be multiplied by a factor of 2.0.

For the Ft. Wayne estimates of chronic exposure, the isopleths were drawn by using "average surface dosages." It appears that the sum, instead of the average, should have been used to plot these isopleths.

For the pathway of direct atmospheric convection and inhalation, Table G-4 summarizes maximal integrated exposures in the 4 cities considered in this report. In Corpus Christi, the maximal integrated concentration for each release was reported at the station nearest the line of release. The maximal integrated-exposure estimate was taken as the sum of the integrated concentrations at that monitoring station.

TABLE G-4 Maximal Integrated Exposures and Lung Deposition of Cadmium by Direct Atmospheric Convection and Inhalation Pathway

City	Area	Maximal Integrated Exposure (particle-min/L)	Cadmium (mg-min/m^3)	Lung Deposition Cadmium (mg)
Corpus Christi		2.3E+02	4.3E-03	7.2E-05
Minneapolis	Able	1.7E+05	3.2E+00	5.3E-02
	Baker	2.8E+04	5.3E-01	8.8E-03
	Charlie	2.1E+04	3.9E-01	6.5E-03
	Dog	8.0E+03	1.5E-01	2.5E-03
	Easy	2.4E+04	4.4E-01	7.4E-03
	Citywide	1.1E+05	2.0E+00	3.3E-02
	Clinton School	2.5E+02	4.6E-03	7.6E-05
Ft. Wayne		1.6E+04	2.9E-01	4.8E-03
St. Louis	How	4.9E+04	9.1E-01	1.5E-02
	Item	1.6E+04	2.9E-01	4.8E-03
	Citywide	1.0E+02	1.9E-03	3.1E-05

For the Minneapolis and St. Louis experiments, the maximal integrated-exposure estimates for each area were found by summing across all releases for each area the values of the isopleth above the highest observed concentrations. For Minneapolis, data from the Army risk-assessment table 7 were used (AEHA 1994b). For example, the "Baker" area had 2 maximums within the 10,000-particle-min/L isopleth, 8 within the 1,000-particle-min/L isopleth, and 4 within the 100 particle-min/L isopleth. Thus, the maximal integrated exposure estimate was taken to be

$$(2 \times 10{,}000) + (8 \times 1{,}000) + (4 \times 100) = 28{,}400 \text{ particle-min/L}.$$

That estimate is conservative in that it uses the isopleth level exceeding

APPENDIX G 327

the highest reported measurement and assumes that the same person was exposed to the maximum from every release within an area. For the St. Louis experiments, the data for the calculations were taken from table 6 of the Army risk assessment (AEHA 1994c).

For the Ft. Wayne experiments, the maximal exposure estimate was calculated by summing the maximal integrated concentration for each release from table 3 in the Army risk assessment (AEHA 1995). The maximal integrated exposure was then converted from particle-minutes per liter to cadmium in milligram-minutes per cubic meter by using the adjustments mentioned above. Using the data from Corpus Christi as an example:

$$232.7 \frac{\text{part-min}}{\text{L}} \cdot \frac{\text{g FP}}{2.16 \times 10^{10} \text{ part}} \cdot \frac{10^3 \text{L}}{\text{m}^3} \cdot \frac{10^3 \text{ mg}}{\text{g}} \cdot \frac{0.156 \text{ mg Cd}}{\text{mg FP}} \cdot \frac{1}{0.39} = 0.00431 \frac{\text{mg Cd-min}}{\text{m}^3}.$$

The integrated exposures in terms of cadmium were then used to estimate the lung deposition of cadmium, assuming 100% deposition and a breathing rate of 1 m³/h, shown below for the Corpus Christi data:

$$0.00431 \frac{\text{mg Cd-min}}{\text{m}^3} \cdot \frac{1 \text{ m}^3}{\text{hr}} \cdot \frac{\text{hr}}{60 \text{ min}} = 7.18 \times 10^{-5} \text{ mg Cd.}$$

ASSESSMENT OF NONCANCER HEALTH EFFECTS

The subcommittee's evaluation of the noncancer health effects addressed in the AEHA report is presented below.

The Army states that because the Environmental Protection Agency did not have a noncarcinogenic-toxicity comparison value for inhalation (a reference concentration, or RfC) for ZnCdS or for cadmium, a number of assumptions, regarding both exposure and toxicity, had to be made in its assessment of risk, namely, that cadmium is a proxy for ZnCdS, that ZnCdS is insoluble, and that inhalation is the exposure pathway. Major factors considered by the subcommittee include the appropriateness of using cadmium-toxicity data in the Army's assessment of the risk associated with exposure to ZnCdS; the assessment of both acute toxicity and chronic toxicity as related to the dispersion tests, and the evaluation of the appropriateness of the conclusions drawn in the Army's report.

The risk assessment developed by the Army correctly emphasizes that it is the solubility of cadmium compounds that eventually will play a determining role in causing pulmonary or systemic effects. ZnCdS is considered to be much less bioavailable and therefore less toxic than many other cadmium compounds. Furthermore, it is pointed out that lung injury caused by inhalation of high concentrations of cadmium compounds, as seen in experimental animals, can range from signs of acute diffuse alveolar damage, with accompanying edema and regenerative changes during the recovery phase, to chronic degenerative lesions. Acute exposure to high concentrations of cadmium have caused death in humans, but no data are known on adverse pulmonary effects in humans exposed chronically to low concentrations. Under such conditions, renal toxicity, and not pulmonary toxicity, might be the adverse health effect that is of most concern.

Because of the lack of toxicity data on the health effects of ZnCdS, it was reasonable to assume, for the purpose of assessing risk, that cadmium sulfide, CdS, would be the most-toxic constituent. The health risk assessment conducted by the Army assumed a worst-case situation regarding the retention and absorption rate of the inhaled test substance. The assumption was that the airborne particles would be so small as to be 100% respirable. Considering the low airborne concentrations of ZnCdS aerosol measured in the Army's dispersion studies and the short exposure period, it is unlikely that such exposure to the substance would be sufficient to cause lung injury.

With regard to the assessment of possible noncarcinogenic pulmonary effects, the subcommittee concurs with the conclusion of the Army that, because of the low concentration and duration of exposure, the ZnCdS tests posed negligible pulmonary-health threats to residents in the test areas.

In the final risk assessment for noncancer effects, the Army document deals with kidney damage, and not with potential lung damage, caused by cadmium and then compares possible exposure to ZnCdS compounds with accepted standards for exposure of workers to cadmium compounds in industry. That approach seems to be justified.

The Army's health risk assessments for Minneapolis, MN, Fort Wayne, IN, and Corpus Christi, TX are adequate in addressing concerns regarding adverse health effects among those environmentally exposed populations. As stated in each of the 4 documents, several assumptions underlie the risk assessments (that cadmium is a proxy for ZnCdS, that ZnCdS is insoluble, and that inhalation is the exposure pathway). The Army states the limitations of its exposure estimates and appears to be conservative in interpreting results, given the many uncertainties associated with exposure estimation. Children were recognized to be more sensitive but no estimates of exposure were made for children except in Minneapolis.

The subcommittee concurs with the conclusions of the Army's health risk assessment for noncancer effects of the dispersion of ZnCdS. Because of the lack of toxicity data on the health effects of this chemical, the Army's assessment was based on available information on exposure to cadmium. That is reasonable because any expected toxicity due to ZnCdS exposure should be related to the cadmium available and not to the relatively nontoxic zinc. The subcommittee agrees that because of the low concentrations and short duration of the exposure, the ZnCdS tests posed negligible pulmonary-health threats to the residents in the test areas. With those considerations, the Army's health risk assessment is reasonable.

The subcommittee notes four concerns. First, there is no description of the population at risk in any document. From an epidemiologic perspective, coupled with appropriate attention given to "sensitive" subpopulations, a description of the population at risk would be helpful. Specifically, census data could be used to describe the population at risk by age, sex, population density, and economic status. The subcommittee's assessment also recognized some uncertainties that exist when applying the standards for healthy workers to the general population. The Army understands that in the general population, some people, such as the elderly and children might be more sensitive to the effects of cadmium than the healthy worker. As a general toxicologic practice, one does not assume that a safe level for adults is protective for children. The Army assumed that because the calculated air concentration was lower than the established occupational safety levels and the exposure was for a short period, the air concentration in the testing area would not represent any undue hazard for the general population. Although the Army recognized that children could be more sensitive, no estimates of exposure were made for children except in Minneapolis, MN.

A second concern stems from the Army's inconsistent use of the terms used in the toxicity-assessment part of the documents. Such terms as "inconclusive studies," "limited data," "not generally associated," and "inconclusive data" are fairly close in meaning but subject to individual interpretation. Operational definitions for terms would be helpful in providing the reader with a sense of whether there are data, whether the data support an association (with or without statistical significance), or whether the studies are of sufficient quality. Regardless of the definitions chosen, consistent application of the terms to each organ system is warranted.

The third concern is that the executive summaries in the 4 documents express reproductive effects in slightly different, albeit subtle, ways. The executive summaries for Corpus Christi and Minneapolis state that there are no adverse reproductive outcomes among women occupationally exposed to higher concentrations of cadmium for longer periods. Later, in the toxicity-assessment sections, the Army notes that reductions in birthweight among offspring born to exposed women were observed in 2 studies. In Fort Wayne, "no deformities in offspring . . . were noted." The summary is correct in noting that reproductive risk has not been fully characterized. A spectrum of potential reproductive or developmental outcomes would need to be included to assess reproductive and developmental toxicity fully, especially at lower (background) levels of exposure.

Finally, the lung is a prime target for airborne substances, but other routes, such as dermal exposure and ingestion, should have been considered. The exposure profiles of adults and children can differ widely. Children tend to have more oral exposure, from hand-to-mouth contact, which could increase the potential for ingestion.

ASSESSMENT OF CANCER RISK

The subcommittee evaluated the cancer risk assessments for Corpus

Christi, TX, Minneapolis, MN, St. Louis, MO, and Fort Wayne, IN, conducted by the Army Environmental Hygiene Agency (AEHA 1994a,b,c, 1995). The general approach used was that of the Environmental Protection Agency, namely, the lifetime cancer risk was calculated as the product of cancer potency and average lifetime daily dose. Because the potency was calculated to represent an upper bound for low-dose cancer-risk estimation,

$$\text{risk} \leq \binom{\text{cancer}}{\text{potency}} \times \binom{\text{average}}{\text{daily dose}}$$

That procedure is widely used and is adopted for this report.

The Army used a cancer potency of inhaled cadmium of 6.3 mg/kg per day on the basis of studies in rodents (Integrated Risk Information System, Office of Health and Environmental Assessment, Environmental Protection Agency, Washington, DC). As discussed in Chapter 5, we have chosen to use a lung-cancer potency estimate of 11.7/(mg/kg) per day based on lung cancer observed in cadmium workers. That increases the Army estimates of risk by a factor of $11.7/6.3 = 1.9$.

Arguments are presented in chapter 5 for increasing the cancer risk estimates by a factor of 10 to account for greater sensitivity at some ages, particularly for children.

The Army used an inhalation rate of air of 0.8 m^3/h. For active people and workers, the National Research Council typically uses an inhalation rate of 1 m^3/h. That increases the dose, and hence the risk, by a factor of $1/0.8 = 1.25$.

The Army used an average lifetime of 70 yr, and the Research Council typically uses 75 yr. That averages out the cadmium dose over a longer period and reduces the average daily dose, and hence the cancer risk, by a factor of $70/75 = 0.93$.

The Army considered cadmium to be 20% of ZnCdS. The subcommittee used a value of 15.6% that reduces the cadmium exposure by a factor of $15.6/20 = 0.78$.

The efficiency of the particle disseminators used in some of the tests was 39%; that is, the weight per particle was $1/0.39 = 2.56$ times as high as that used by the Army. Thus, the concentrations and risks need to be multiplied by 2.56.

The combination of the 6 factors makes the subcommittee cancer risk estimates higher than the Army estimates by a factor of

$$1.9 \times 10 \times 1.25 \times 0.93 \times 0.78 \times 2.56 = 44.$$

The major difference is the factor of 10 introduced to allow for the possible additional sensitivity of particular age groups.

It appears that the Army estimates of cancer risk are based on the maximum average exposure at a site for 1 release in the Fort Wayne tests. There were 35 releases each of green and yellow cadmium compounds in the Ft. Wayne tests, so the cancer risk estimate should be increased by a factor of 35. The disseminator efficiency for the Ft. Wayne tests was 33%, compared with 39% at the other sites, so the overall modifying factor should be increased to $44 \times (0.39/0.33) = 52$, increasing the Ft. Wayne estimates by a factor of $35 \times 52 = 1,820$. Even with appropriate adjustments to the cancer risks calculated by the Army, the lifetime lung-cancer risks associated with the low doses of cadmium are low, and it is unlikely that anyone in these 4 test areas developed lung cancer as a result of direct inhalation of cadmium from the airborne releases.

REFERENCES

AEHA (U.S. Army Environmental Hygiene Agency). 1994a. Assessment of Health Risk, Corpus Christi, Texas. HRAS No. 64-50-93QE-94. U.S. Army Environmental Hygiene Agency, Aberdeen Proving Ground, Edgewood, Md.

AEHA (U.S. Army Environmental Hygiene Agency). 1994b. Assessment of Health Risk, Minneapolis, Minnesota. HRAS No. 64-50-93QE-94. U.S. Army Environmental Hygiene Agency, Aberdeen Proving Ground, Edgewood, Md.

AEHA (U.S. Army Environmental Hygiene Agency). 1994c. Assessment of Health Risk, St. Louis, Missouri. HRAS No. 64-50-93QE-94. U.S. Army Environmental Hygiene Agency, Aberdeen Proving Ground, Edgewood, Md.

AEHA (U.S. Army Environmental Hygiene Agency). 1995. Assessment of Health Risk, Fort Wayne, Indiana. HRAS No. 39-26-0467-95. U.S. Army Environmental Hygiene Agency, Aberdeen Proving Ground, Edgewood, Md.

Hilst, G.R., and N.E. Bowne. 1966. A Study of the Diffusion of Aerosols Released from Aerial Line Sources Upwind of an Urban Complex. Vol. 1, Final Report; Vol. 2, Data Supplement. Contract DA-42-007-AMC-37(R). Prepared by the Travellers Research Center, 250 Constitution Plaza, Hartford, Conn., for the U.S. Army Dugway Proving Ground, Dugway, Utah.

NRC (National Research Council). 1988. Emergency and Continuous Exposure Guidance Levels for Selected Airborne Contaminants, Volume 8, Litium Chromate and Trichloroethylene. Washington, D.C.: National Academy Press.

Smith, T.B., and M.A. Wolf. 1963. Vertical Diffusion from an Elevated Line Source over a Variety of Terrains. Part A. Final Report. Contract DA-42-007-CML-545. Prepared by Meteorology Research, 2420 North Lake Ave., Altadena, Calif., for the U.S. Army Dugway Proving Ground, Dugway, Utah.

Appendix H

Review of EPA, ATSDR, and CDC Comments on the Army's Risk-Assessment Reports on Zinc Cadmium Sulfide

Review of EPA, ATSDR, and CDC Comments on the Army's Risk-Assessment Reports on Zinc Cadmium Sulfide

THE U.S. ARMY REQUESTED that the U.S. Environmental Protection Agency (EPA), the Centers for Disease Control and Prevention (CDC), and the Agency for Toxic Substances and Disease Registry (ATSDR) review its health risk-assessment reports for the tests conducted with ZnCdS.

The Subcommittee on Zinc Cadmium Sulfide had the opportunity to review comments by EPA, CDC, and ATSDR experts on the Army's risk assessments for releases of ZnCdS in Corpus Christi, TX, and Minneapolis, MN. A total of 7 EPA experts wrote comments, which were summarized by Hugh McKinnon (director of EPA's Health Assessment Group). Comments from experts at the CDC and ATSDR were also available.

EPA Review

EPA reviewed the Army Environmental Health Agency's risk-assessment reports for Corpus Christi, TX, and Minneapolis, MN. EPA indicated general agreement with the Army's conclusion, stating that "if the exposure calculations are even close to correct (that is, within a hundredth to a millionth of the actual exposure), the test sites appear to pose no human health risk." However, EPA suggested several possible improvements in the documents. As stated by Robert Huggett, assistant administrator for research and development at EPA, "We found the Army's conclusion of no health threat is likely correct but too narrowly stated."

EPA suggested that the Army's assessment documents discuss exposures of sensitive subpopulations, such as children and debilitated adults. There was also concern that inhalation was the only exposure route considered and that neglect of the dermal and oral routes led to underestimation of the cadmium exposure. EPA concluded that some assumptions made can result in underestimates of exposure, perhaps by wide margins, and recommended that the margins of uncertainty be presented, in addition to the estimates themselves. EPA suggested that the Army's assessment of exposure downwind of the dispersion devices did not represent the maximal ZnCdS concentrations, which would be upwind of the monitored areas (closer to the dispersion device).

ATSDR AND CDC REVIEWS

ATSDR's and CDC's reviews of the risk associated with ZnCdS dispersion tests concluded that the tests would result in negligible increases in background concentrations of cadmium and zinc in air, soil, and water. The ATSDR and CDC reviewers concurred with the conclusion of the Army that the ZnCdS tests posed negligible health threats to residents of the Corpus Christi and Minneapolis test areas.

In general, the reviewers did not deal in depth with the methodology of measurement, computation of exposures, or translation of exposures to health risks. In addition, most did not comment on the lack of attention to risks other than cancer in the Army's documents.

Some reservations were expressed regarding routes of exposure other than inhalation, possible bioaccumulation, risk to special populations, and margins of uncertainty, but the reviewers' comments generally were not detailed.

Appendix I

Cadmium Exposure Assessment, Transport, and Environment Fate

Cadmium Exposure Assessment, Transport, and Environment Fate

THE PURPOSE OF THIS APPENDIX is to provide estimates of the magnitude of potential human contact with cadmium compounds as a result of the dispersion of zinc cadmium sulfide, ZnCdS, by the Army.

EXPOSURE TO ENVIRONMENTAL CADMIUM

Cadmium is a chemical element and a natural component of the earth's crust. Human activities can increase human exposure to cadmium through mining and combustion, which bring more cadmium into the air, water, and soil. In the sections below, we summarize sources of human exposure to cadmium in air, water, plants, animals, food, soil, and house dust. For each, we characterize likely magnitudes of human contact and the routes of contact—inhalation, ingestion, and dermal contact.

Cadmium in Outdoor Air

Cadmium metal and cadmium salts have low volatility and exist in air primarily as fine suspended particulate matter. It enters the air from burning coal and household wastes, and from metal mining and refining processes. In the United States, mean levels of cadmium in ambient air range from less than 0.001 $\mu g/m^3$ in remote areas to 0.005-0.04 $\mu g/m^3$ in urban areas (Davidson and others 1985; Elinder 1985; EPA 1981; Saltzman and others 1985). Atmospheric concentrations of cadmium are generally highest in the vicinity of cadmium-emitting industries such as smelters, municipal incinerators, or fossil fuel combustion facilities. Measurements of atmospheric cadmium up to 7 µg/m^3 have been reported in these industrial types of areas in the United States (Schroeder and others 1987). When inhaled, some fraction of this particulate matter is deposited in the airways or lungs, and the rest is exhaled. Because most people spend only about 10% or less time outdoors per day, this issue is addressed in the exposure analysis. In the United States, a person who breathes 20 m^3 of air per day and spends 10% of his or her time outdoors will have an estimated cadmium intake of 0.1-0.8 µg/day in urban cities or less than 0.02 µg/day in rural areas.

Cadmium in Water

Cadmium enters drinking water directly from pollution-source releases to surface water and groundwater or from deposition from air to surface water, from soil runoff to surface water, or from leaching from rocks and soils into groundwater. The concentration of cadmium dissolved in the open ocean is less than 0.005 µg/L (IARC 1993; Nriagu 1980). The concentration of cadmium in drinking water is generally reported to be less than 1 µg/L, but it might increase to 10 µg/L as a result of industrial discharge and leaching from metal and plastic pipes (Friberg and others 1974). A person who consumes 2 L of water daily with a cadmium concentration of 1 µg/L will have an intake of 2 µg/d.

Uptake in Plants and Cadmium in Food

Plants are contaminated with cadmium via two routes—uptake of cadmium in soil through the roots and deposition of cadmium in air onto leaf surfaces followed by translocation to other plant parts. Cadmium residues in plants are typically less than 1 µg/kg (IARC 1993). Food is the main source of cadmium for nonoccupationally exposed people, although uptake of cadmium in the gut from food is generally less efficient than from water or air because cadmium binds to food constituents (IARC 1993). The average daily intake of cadmium through food varies among countries, and varies among individuals in a given country or population group. In nonindustrialized rural areas away from mining operations, cadmium intake through food is estimated to be 10-60 µg/d; in polluted areas of Japan, values as high as 500 µg/d have been reported (Friberg and others 1974).

There are several estimates of the daily adult intake of cadmium from food in the United States, but there is considerable variation among those estimates. Schroeder and Balassa (1961) reported a range of 4-60 µg per day and Nriagu (1981) reported a range of 38-92 µg, while estimated daily averages have been reported to be 30 µg (Gartrell and others, 1986), 38 µg (Duggan and Corneliussen, 1972), 50 µg (Duggan and Corneliussen, 1972), 51 µg (Mahaffey and others, 1975), and 92 µg (Murthy and others, 1971). A more recent estimate based on a Total Diet Study shows the daily dietary intake to be approximately 15 µg (Gunderson, 1995). Analysis of the earlier data shows that these discrepancies are probably due to different analytical methods. Cadmium contamination of food has been reduced over the years, presumably because of better technology. However, the cadmium contamination encountered in the 1950s and 1960s when the Army's dispersion tests were conducted are more relevant for risk assessment. On the basis of the U.S. data and data from other industrial nations from the northern hemisphere, the subcommittee believes that the daily cadmium intake from food ranges from 10 to 60 µg.

CADMIUM IN SOIL (SOIL INGESTION AND DERMAL UPTAKE)

Human intake of cadmium in soil occurs through soil ingestion that results from hand-to-mouth activities. Such intake is typically less important than the inhalation, water-intake, and food-consumption pathways associated with the same soil (McKone and Daniels 1991). Cadmium concentrations in soil vary widely. In nonpolluted areas, they are usually below 1 µg/g; in polluted areas, concentrations of up to 800 µg/g have been detected (Friberg and others 1974).

The U.S. Environmental Protection Agency (EPA 1992) recommends that 0.2 g/d be used as an average soil-ingestion rate for children under age 7 with 0.8 g/d as an upper bound. LaGoy (1987) suggests a soil-ingestion rate of 0.025 g/d for adults. On the basis of those values, we estimate that, in an area with soil cadmium of around 1 µg/g, cadmium intake as a result of soil ingestion is in the range of 0.02-0.2 µg/d.

According to EPA (1992), an adult with a body surface area of 18,000 cm^2 can have 5,000-5,800 cm^2 of skin area exposed to soil contact and have a soil-to-skin adherence of 0.2-1 mg/cm^2 with an exposure frequency of 40-350 events per year. That translates into an equivalent annual soil contact of 0.1-6 g/d. No human data on dermal uptake of cadmium are available, but animal data indicate that dermal uptake is slow, about 2% over 24 h of contact (ATSDR 1993). On the basis of these values, human contact with soil containing cadmium at 1 µg/g would result in annual dermal uptake of 0.002-0.12 µg/d.

Cadmium in House Dust

In recent years, it has been recognized that fine and coarse particles in the indoor environment have both air and soil sources and enter the indoor environment by such processes as resuspension, deposition, and soil tracking (Allott and others 1992; Nazaroff and Cass 1989). Friberg and others (1974) reported that the concentration of cadmium in the dust deposited in houses was related to concentrations in air particles more than to soil concentrations. In rural areas with cadmium concentrations in air of 0.005 $µg/m^3$, cadmium was contributed to the house-dust pool at around 600 m/d. That implies that the cadmium deposition rate on household surfaces is about 3 $µg/m^2$ per day in areas with an air concentration of 0.005 $µg/m^3$. The resulting house-dust cadmium concentrations were in the range 13-14 µg/g of dust (Friberg and others 1974). That is much higher than the 1 µg/g typical of soil in the same areas and suggests the relative importance of deposition from air to house dust as a potential exposure pathway. The ratio of dust to air concentration suggests that cadmium concentrations could be from around 40 µg/g of dust in urban areas to as much as 140 µg/g of dust in industrialized areas, on the basis of air concentrations in these areas reported above. However, it should be noted that ambient air is not the only likely source of cadmium in indoor air. Indoor sources, such as smoking and the agents used to color carpets and furniture, could result in increased cadmium in house dust.

The loading of soil and dust on floors as reported in the recent literature varies from 0.136 to 0.870 g/m^2 (Allott and others 1992; Nazaroff and Cass 1989). If we assume a dust inventory of 1 g/m^2 on household surfaces, then, according to the air concentrations discussed above, the inventory of cadmium in the dust of household surfaces is about 14 μg/m^2 in rural areas, 42 μg/m^2 in urban areas, and as much as 140 μg/m^2 in industrial areas. The residence time of cadmium in this dust is around 5 d; this is obtained by dividing the cadmium inventory, such as 14 μg/m^2, by the rate of cadmium deposition on household surfaces, 3 μg/m^2 per day.

The soil ingestion and soil dermal-uptake calculations developed above for outdoor soils indicate that for soil ingestion the intake-to-concentration ratio is about 0.02-0.01 μg of cadmium ingested per day per 1 μg/g of soil and that for dermal contact the uptake-to-concentration ratio is about 0.002-0.12 μg of cadmium dermally adsorbed per day per 1 μg/g of soil. Because dust contacts indoors are likely to be lower than soil contact outdoors, consider the lower bound on these ranges to be appropriate for house dusts. In this case, the combined uptake from the two exposure routes is about 0.02 μg of cadmium per day per 1 μg/g of soil. If this ratio were applied to house dusts, the average ingestion and dermal uptake of cadmium from house dust would both be about 0.3 μg/d in rural areas, 0.8 μg/d in urban areas, and as much as 3 μg/d in industrial areas.

Cadmium in Indoor Air

In assessing the concentration of cadmium in indoor air, it is important to consider both outdoor air and house dust as sources. As noted above, cadmium concentrations in house dust are typically about 15 μg/g in rural areas, 40 μg/g in urban areas, and 140 μg/g in industrial areas. Assuming that the indoor air of the house has a dust loading of 50 μg/m^3 and that that would be the same in household air would yield respective indoor air concentrations of 0.007 μg/m^3 in rural areas, 0.02 μg/m^3 in urban areas, and 0.07 μg/m^3 in industrial areas. Given that these levels are comparable with those of outdoor air, resuspended dust inside homes is not likely to increase cadmium concentrations in indoor air substantially. Instead, the figures suggest that cadmium enters the indoor air mainly from outside and not from dust resuspension.

SMOKING

It has been recognized for many years that smoking can be an important source of cadmium exposure of smokers. Friberg and others (1974) summarized a number of studies relating cadmium uptake to smoking. They reported that all the data agree well and show that 0.1-0.2 μg of cadmium can be inhaled by smoking 1 cigarette. Smoking habits to some extent affect the amount of cadmium taken up. However, it can be estimated that someone smoking 20-40 cigarettes per day will take in 2-8 μg of cadmium per day.

SUMMARY

Exposures to ambient cadmium result in a daily human intake in the range of 12-84 μg/d for an adult. For a 70-kg person, that corresponds to a potential dosage of about 0.2-1.2 μg/kg per day. The relative contributions of the various pathways to total potential dose for a 70-kg adult are as follows:

Inhalation of cadmium in air indoors and outdoors	0.002-0.02 μg/d	~0.02%
Water ingestion	2-20 μg/d	~20%
Food products	10-60 μg/d	~76%
Soil ingestion	0.02-0.2 μg/d	~0.2%
Dermal contact	0.002-0.12 μg/d	~0.002%
Contact with house dust	0.3-3 μg/d	~3%
Smoking	(2-8 μg/d)	(for smokers)
Total daily intake	12-84 μg/d	

Food products and water contribute almost all the typical daily human exposure. Inhalation exposures contribute a very small fraction. From our estimates here, contact with house dust is the third-ranking (for nonsmokers) potential exposure pathway for existing cadmium sources.

Transport and Environmental Fate

This section describes the sources, transport, environmental fate, and accumulation of ZnCdS and cadmium compounds in the environment. We begin by considering the sources and chemical properties of ZnCdS and cadmium compounds. Because ZnCdS does not occur naturally and little has been published on its environmental behavior, little can be reported here. For cadmium compounds, there is a rich, but still incomplete, literature on chemical properties, environmental concentrations, transport, and the global chemical cycle. Some relevant components of that literature are reported here.

Distribution Coefficients in Soils and Sediments

The distribution or sorption coefficient, K_d, is ratio of the concentration, at equilibrium, of a chemical species attached to solids or particles (in moles per kilogram) to chemical concentration in the solution, with which the particles have contact (in moles per liter). Several mechanisms define this partition relationship—including cation exchange, adsorption, speciation, coprecipitation, and organic complexation. Soil-water distribution coefficients are often modeled as independent of water-phase concentration, whereas in reality there often is a dependence of K_d on water-phase concentration. Bodek and others (1988) have reviewed and compared a number of sorption models for cadmium in soil-water and sediment-water systems. They report that in soils estimated K_d values range from 1 to 9,000 with a typical value (at low water concentrations) of about 1,000 and that in sediments estimated K_d values range from 1 to 160,000 with a typical value (at low water concentrations) of about 6,000.

Bioconcentration Factors for Plant Uptake from Soil

The plant-soil partition coefficient, K_{ps}, expresses the ratio of contaminant concentration in plant parts, both pasture and food, (in micrograms per gram of plant fresh mass) to the concentration in wet root-zone soil (in micrograms per gram). Cadmium is considered a potential essential trace element for plants and animals (Mertz 1981). Root uptake of cadmium as Cd^{2+} in plants is passive and occurs though uptake by roots of cadmium dissolved in water; cadmium is highly mobile in plants and readily translocated to other plant parts (Bodek and others 1988). Plant-soil partition coefficients have been reported in the range of 0.015-2.1 with a likely value of about 0.1 (Baes and others 1984; Bowen 1979; Nriagu 1980).

Bioconcentration Factors for Plant-Leaf Concentration Relative to Air Concentration

According to Bodek and others (1988), airborne deposition is believed to contribute to concentrations of cadmium found in plant leaves. At low concentrations, the ratio of plant-leaf concentration to air concentration when air and plant environments are in contact can be estimated as

$$C_p/C_a = v_d/(M_p \times R_p),$$

where C_p is cadmium concentration in the plants in contact with contaminated air, mol/kg; C_a is the cadmium concentration in air above the plants, mol/m³; v_d is the deposition velocity that represents the rate of cadmium transfer from air to plant surfaces, m/d; M_p is the mass of the plants per unit area of land, kg/m²; and R_p is the first-order rate constant that accounts for all cadmium removals from plants (wash-off, biodegradation, and so on) per day. McKone and Ryan (1989) have estimated that, on the basis of the balance between deposition, weathering, and senescence processes in agricultural landscapes, this ratio can be estimated to have a mean value of 3,300 m³ (air)/kg (plants) and a geometric standard deviation of 3 for metal species.

BIOCONCENTRATION FACTORS FOR FISH

The bioconcentration factor (BCF) is a measure of chemical partitioning between fish tissue based on chemical concentration in water and has a unit of moles per kilogram of fish per mol per liter of water. Bodek and others (1988) report ocean and freshwater fish BCFs in the range of 200-50,000, with 2,000 being a typical value in this range of reported values.

SOURCES AND SINKS

It is not clear from the current geochemical literature whether $Zn_{0.8}Cd_{0.2}S$ occurs naturally and has its own biogeochemical cycle. In contrast, there has been extensive study of the natural and human biogeochemical cycles of cadmium (for example, Nriagu 1980). For $Zn_{0.8}Cd_{0.2}S$, our principal concern is the potential exposure to cadmium. That requires that we understand the local and regional scale distribution and fate of both $Zn_{0.8}Cd_{0.2}S$ and the cadmium compounds formed from it. In the areas of small geographic extent to which it was dispersed, the persistence of cadmium will depend on the rate of atmospheric dispersion, the rate of deposition to surfaces (soil, snow, plants, and so on), and the rate of transformation from $Zn_{0.8}Cd_{0.2}S$ to some other cadmium compound. Before the specific behavior of $Zn_{0.8}Cd_{0.2}S$ in the environment can be fully characterized, there is a need to determine both the rates and end products of $Zn_{0.8}Cd_{0.2}S$ transformation reactions in air, surface soil, vegetation surfaces, surface water (rivers, lakes, and ponds) and snow. Some possible reactions are listed below, but these are difficult to confirm.

$$5\ Zn_{0.8}Cd_{0.2}S + 10\ H_2O + 5\ O_2 \longrightarrow 4\ Zn^{2+} + Cd^{2+} + 5\ SO_4^- + 10\ H^+$$
$$5\ Zn_{0.8}Cd_{0.2}S + 10\ O_2 + \text{photon energy} \longrightarrow 4\ Zn^{2+} + Cd^{2+} + 5\ SO_4^-$$
$$Zn_{0.8}Cd_{0.2}S + \text{biologic organisms} \longrightarrow Zn^{2+}? + Cd^{2+}? + ?$$

Although the actual degradation rate of $Zn_{0.8}Cd_{0.2}S$ has not been measured in any environmental medium, the degradation of the fluorescence of $Zn_{0.8}Cd_{0.2}S$ has been studied (Leighton and others 1965) and is likely related to the rate of transformation of $Zn_{0.8}Cd_{0.2}S$ to other zinc and cadmium compounds. The degradation of $Zn_{0.8}Cd_{0.2}S$ fluorescence has been attributed to photochemical reactions and found to vary with humidity and manufacturing lot (Leighton and others 1965). One study reported a loss of as much as 50% of $Zn_{0.8}Cd_{0.2}S$ particles within 2 h of airborne travel in sunlight (Eggleton and Thompson 1961). Using the type of $Zn_{0.8}Cd_{0.2}S$ used in several Army studies, Leighton and others (1965) reported only a 7% loss after a period of 19 h that included mostly daylight hours.

The sources, sinks, and distribution of cadmium in many ecosystems have yet to be properly evaluated, and cadmium transfer rates between components of the earth system are only poorly known. Aspects of the global cycle of cadmium have been summarized by Nriagu (1980). Basically, a global model of an element consists of presumed reservoirs or compartments (atmosphere, lithosphere, ocean, lakes, soils, and so on), which can be active (available to the biota) or passive (unavailable to the biota). The total amount of cadmium in each compartment is referred to as the burden (or pool) and is obtained by multiplying the average concentration by the total mass (or volume) of the compartment. Reactions and advection processes in compartments can result in smaller-scale variations of cadmium concentration with both space and time.

Table I-1 shows the principal compartments and concentrations of cadmium in the surface environment of the earth; the numbers are taken from Nriagu (1980). It should be noted that active pools—such as the atmosphere, soils, lakes, rivers, and ocean pools—are subject to large inputs of cadmium from human activities. The exchange of cadmium between compartments occurs along established pathways involving stream, ice, and groundwater flows; atmospheric transport and deposition; volcanism; uplift; weathering; sedimentation; and biologic mobilization. The residence time of each compartment gives a sense of how long that compartment takes to respond to a change in cadmium inventory. From the table we can conclude that the major natural sources of cadmium to the active parts of the environment are mobilization from the large reservoir that exists in the lithosphere. The major sink for cadmium that enters the active compartments is burial in freshwater or ocean sediments.

TABLE I-1 Cadmium (Cd) Burdens and Residence Times in the Principal Global Compartments as Reported by Nriagu (1980)

Compartment	Cadmium Concentration	Total Cadmium in Compartment (g)	Residence Time (d)[a]
Atmosphere	0.03 ng/m^3	1.5×10^8	7
Hydrosphere:			
Ocean	0.06 µg/kg	9×10^{13}	9×10^6
Surface fresh water	0.05 µg/kg	2×10^9	na
Fresh water Sediment layer	0.16 µg/g	1×10^{11}	1,300
Groundwater	0.1 µg/kg	4×10^{13}	na
Sediment-pore water	0.2 µg/kg	6×10^{13}	na
Glaciers	0.005 µg/kg	8×10^{10}	na
Biosphere:			
Marine biota	2 µg/g	1×10^{10}	20
Land plants	0.3 µg/g	7×10^{11}	20
Land animals	0.3 µg/g	6×10^9	na
Fresh-water biota	3.5 µg/g	7×10^9	4
Soil Surface-soil litter and detritus to 1 cm	0.6 µg/g	1×10^{12}	15,000
Root soil to 1 m	0.2 µg/g	7×10^{13}	1×10^6
Lithosphere—sedimentary rock to 45 km	0.5 µg/g	3×10^{19}	1×10^9

[a]na = not available.

EXPOSURE ASSESSMENT

Assessment of human exposure to cadmium requires translating environmental concentrations and sources into estimates of the amount of cadmium that comes into contact with the population at risk. This contact is the basis for estimating potential doses used in health risk assessments. Potential dose, expressed as average daily dose, is the amount of material per unit of body weight per day (milligrams per kilogram per day) that enters the lungs (inhalation route), enters the gastrointestinal tract (ingestion route), or crosses into the stratum corneum (dermal route). The total potential dose is used as a basis for projecting the incidence of health effects within the population.

In estimating the exposures to cadmium in areas where $Zn_{0.8}Cd_{0.2}S$ was dispersed by the Army, it is important to assess not only the amount of material dispersed and the resulting short-term air concentrations, but also the extent to which this material was deposited on soil, vegetation, and snow and tracked into houses and other buildings. We explore here the premise that the effective persistence of the exposure concentrations in surface soil, house dust, and vegetation can be longer than those in the atmosphere, where cadmium persistence is measured in hours. Material transferred to soil surfaces, house dust, or plants could persist as an increased source of exposure for longer periods, perhaps days or weeks or months.

In the sections below the potential exposure to cadmium in air is calculated for both the direct (inhalation) exposure pathway and the indirect exposure pathways, including dust resuspension, house-dust contacts, and deposition on vegetation used for food. For each pathway, we develop the potential dose ratio (Table I-2), which is the ratio of cumulative intake for an exposure event, in micrograms of cadmium, divided by the overall product of concentration and time—the time integral of the exposure concentrations in air.

DIRECT EXPOSURE: INHALATION OF CONTAMINATED AIR

To determine the potential dose ratio, we first determine dose, which is the product of inhalation rate, IR, m^3/h; the time-averaged cadmium air concentration during the exposure time, C_{air}, $\mu g/m^3$; and the exposure time, ET, in h. The dose ratio is obtained by dividing the dose by the cumulative exposure, which is the product of average cadmium concentration in air, C_{air}, $\mu g/m^3$, during the exposure time and the exposure time, ET. For direct exposure by inhalation, the potential dose ratio is obtained as follows:

potential dose ratio (inhalation) = $(IR \times ET \times C_{air})/(C_{air} \times ET)$ = IR.

For an adult breathing air at a rate of 1 m³/h,

potential dose ratio (inhalation) = 1 μg/(μg/m³)-h.

When this dose ratio is multiplied by the cumulative exposure for any exposure location, we obtain an estimate of inhalation dose for a person at that location for the period ET.

TABLE I-2 Summary of Potential Dose Ratios for Direct and Indirect Exposure Pathways[a]

Exposure Pathway	Direct or Indirect Contact with Air	Potential Dose Ratio, μg ÷ μg-h/m³ of Cumulative Air Exposure
Inhalation	Direct	1.0
Inhalation of resuspended soil outdoors	Indirect	0.0005-0.002
Dermal contact with and ingestion of house dust	Indirect	2.5-5
Inhalation of resuspended house dust	Indirect	≈ 0.12
Deposition on vegetation in home gardens	Indirect	2.2
Deposition on surface drinking-water supplies	Indirect	0.24-0.8

[a]Potential dose ratio is the ratio of total uptake or intake of cadmium in micrograms, divided by the integrated air exposure, in micrograms per cubic meter per hour.

INHALATION OF PARTICLES RESUSPENDED IN OUTDOOR AIR

The Army risk assessment did not consider exposures to cadmium that was deposited on the ground and then resuspended in the air, where it could be inhaled after the main source had dispersed. The concentration of cadmium added to soil as a result of deposition from air can be estimated as

$$C_{soil} = (C_{air} \times Vd_o \times ET)/(M_{soil}),$$

where C_{soil} is the concentration of cadmium added to soil, µg/g; Vd_o is the outdoor deposition velocity of FP, assumed to be 20 m/h; C_{air} is the cadmium concentration in air during an event, µg/m^3; ET is the exposure time, h; and M_{soil} is the mass inventory of surface soil, which, on the basis of soil depth of 1 cm and a soil density of 1,600 kg/m^3, is 16,000 g/m^2. Combining these values gives

$$C_{soil} \approx 0.00125 \, (C_{air} \times ET) \text{ µg/g}.$$

Soil erosion rates in the United States due to wind and water are around 100 g/m^2 per year. That implies that the residence time of soil particles deposited onto the soil surface is about 160 yr, that is, 16,000 g/m^2 divided by 100 g/m^2-yr. However, because of its relatively high solubility in water, cadmium species are eroded more rapidly from soil than is the bulk soil. Thus, cadmium has a residence time in soil that is more like 10-40 yr (Nriagu 1980).

In urban areas, the concentration of particulate matter in air is about 100 µg/m^3. If resuspended surface soil is assumed to be the source of all the particulate matter in air, the long-term concentration of cadmium in air resuspended from soil is given by

$$\begin{aligned}C_{resuspend} &= C_{soil} \times 100 \text{ µg/m}^3 \times 10^{-6} \text{ g/µg} \\ &= 1.25 \times 10^{-7} \times (C_{air} \times ET) \text{ µg/m}^3.\end{aligned}$$

For someone who breaths this at a rate of 20 m^3/d for 10-40 yr, or 3,650-

14,600 days (the assumed residence time in soil), the potential dose is

$$\text{potential dose} \approx C_{resuspend} \times (3{,}650\text{-}14{,}600 \text{ d}) = (0.0005\text{-}0.002) \times (C_{air} \times ET) \text{ μg,}$$

and thus the potential dose ratio for resuspension is estimated as

$$\text{potential dose ratio (inhalation)} \approx (0.0005\text{-}0.002) \text{ μg/(μg/m}^3\text{)-h [over 10-40 yr].}$$

That value is low relative to the direct exposure by inhalation of contaminated air during the exposure event and would apply only to people who lived in the contaminated area for a number of years after exposure.

Dermal Contact and Ingestion of House Dust

It is difficult to estimate the house-dust impact because we do not have data on penetration and retention within the house. However, we can make a rough estimate of the likely levels of cadmium in house dust and the potential doses based on the existing information on house-dust exposure discussed previously. To estimate dermal and ingestion exposures to the cadmium transported to house dust, it is necessary to determine the cadmium concentration inside houses after dispersion and how much of this is deposited on household surfaces. To make this preliminary calculation, we assume that air concentrations of cadmium in houses went up in proportion to outdoor concentrations during the exposure time, ET, but that house dust is only 50% attributable to outdoor air. On the basis of that assumption, the added concentration of cadmium in house dust is calculated as

$$C_{dust} \approx (0.5 C_{air} \times Vd_h \times ET)/(M_{dust}),$$

where C_{dust} is the concentration of cadmium added to house dust, μ/g; $0.5 C_{air}$ is the assumed cadmium concentration in indoor air during an event, μg/m^3; Vd_h is the indoor deposition velocity of ZnCdS on household surfaces, assumed as discussed above to be 25 m/h (600 m/d); ET is the exposure time, h; and M_{dust} is the mass inventory of dust on household surfaces, assumed as noted above to be about 0.5 g/m². Thus,

$$C_{dust} \approx 25 \, (C_{air} \times ET) \, \mu g/g(dust).$$

House dust has a likely residence time of 5 d (derived earlier). From the increased concentration of cadmium in house dust for 5 d and the previously derived dermal and ingestion uptake factor for house dust of 0.02 µg/d per 1 µg of cadmium per gram of dust, the 5-d effective dose for a single event of duration ET is

$$\text{potential dose} \approx C_{dust} \times 5 \, d \times 0.02 \, g(dust)/d = 2.5 \times (C_{air} \times ET) \, \mu g,$$

so the potential dose ratio for dermal and ingestion contact with house dust is estimated as

$$\text{potential dose ratio (house dust)} \approx 2.5 \, \mu g/(\mu g/m^3)\text{-h [over} \approx 5 \, d].$$

That value is comparable with and slightly higher than the direct exposure to contaminated air by inhalation during the exposure event. However, there is much greater uncertainty associated with it than with the direct inhalation-dose estimate.

INHALATION OF RESUSPENDED HOUSE DUST

To make a preliminary estimate of the potential dose, we assume that air concentrations of cadmium in houses due to house dust are all attributable to resuspension of house dust and that the particle load indoors is around 50 µg/m³.

$$C_{dust \, to \, indoor \, air} \approx C_{dust} \, (\mu g/g) \times 50 \times 10^{-6} \, g/m^3,$$

where C_{dust} is the concentration of cadmium in house dust, µg/g. Because the breathing rate of occupants during this time is around 20 m³/d and the estimated addition of cadmium to house dust during an event of duration ET is

$$C_{dust} \approx 25 \, (C_{air} \times ET) \, \mu g/g(dust),$$

the potential dose to occupants from breathing resuspended house dust is

$$\begin{aligned} \text{potential dose} &\approx 25 \, (C_{air} \times ET) \times 50 \times 10^{-6} \, g/m^3 \times 20 \, m^3/d \times 5 \, d \\ &\approx 0.12 \times (C_{air} \times ET) \, \mu g \\ &\approx 0.12 \, \mu g/(\mu g/m^3)\text{-h [over} \approx 5 \, d]. \end{aligned}$$

That value is much lower than those for the other potential contacts with house dust.

Indirect Exposure from Consumption of Plants and Animals

In theory, plants can be contaminated with atmospheric cadmium either by uptake of cadmium from contaminated soil or by deposition of cadmium from air on leaf surfaces. As noted above, it is difficult to construct a scenario whereby cadmium in soil could be increased by only a small fraction above existing concentrations. Plant roots penetrate deeper into soil than the 1 cm assumed above, so there is no plausible mechanism by which plant tissues could be substantially contaminated by a short-term atmospheric exposure. However, when we consider deposition on leaf surfaces, there is a greater likelihood of vegetation contamination. The concentration of cadmium added to vegetation as a result of deposition from air can be estimated as

$$C_{veg} = (C_{air} \times Vd_v \times ET)/(M_{veg}),$$

where C_{veg} is the concentration of cadmium added to vegetation during an exposure event, $\mu g/kg$; C_{air} is the average cadmium concentration in air during an event, $\mu g/m^3$; Vd_v is the deposition velocity of FP from air to vegetation, assumed to be 10 m/h; ET is the exposure time, h; and M_{veg} is the average density of growing vegetation, which is about 3 kg/m². Combining those values yields

$$C_{veg} \approx 3.3 \, (C_{air} \times ET) \, \mu g/L.$$

On the basis of data from Nriagu (1980), we estimate that particles deposited on vegetation could persist for some 20 d. The important question regarding exposure to vegetation is how much consumption of vegetation would take place among the local populations. EPA data suggest that in urban areas, less than 10% of consumed vegetation is homegrown. Combining that 10% factor, cadmium persistence on plants, and the observation that a typical adult consumes fruits and vegetables at roughly 125 kg/yr, or 0.34 kg/d, we obtain the following estimate of potential dose:

$$\text{potential dose} \approx C_{veg} \times 20 \, d \times 0.10 \times 0.34 \, kg/d = 2.2 \times (C_{air} \times ET) \, \mu g.$$

Thus, the potential dose ratio for ingestion of homegrown vegetables is

$$\text{potential dose ratio (homegrown foods)} \approx 2.2 \, \mu g/(\mu g/m^3)\text{-h [over} \approx 20 \, d].$$

That value is comparable with and higher than the direct exposure to contaminated air by inhalation during the exposure event. However, there is much greater uncertainty associated with it than with the direct inhalation-dose estimate.

Our estimate of exposures through food is likely to be a high-end estimate, because we ignore seasonal effects, we ignore removal of cadmium from plant surfaces by rain and wind, and we assume a relatively high fraction of homegrown food consumption.

INDIRECT EXPOSURE FROM DEPOSITION ON SURFACE WATERS

The concentration of cadmium added to surface water as a result of deposition from air can be estimated as

$$C_{water} = (C_{air} \times Vd_o \times ET)/(V_{water}),$$

where C_{water} is the concentration of cadmium added to surface water during an exposure event, µg/L; C_{air} is the average cadmium concentration in air during an event, µg/m^3; Vd_o is the outdoor deposition velocity of FP, assumed to be 20 m/h; ET is the exposure time, h; and V_{water} is the mass inventory of surface water, L/m^2. On the basis of data in the Water Encyclopedia, we estimate the average depth of surface water as 5 m, which gives a typical value of V_{water} of 5,000 L/m^2. Combining these values yields

$$C_{water} \approx 0.004\ (C_{air} \times ET)\ \mu g/L.$$

The residence time of surface waters—such as lakes, ponds, and rivers—ranges from days to a year. Assuming that added cadmium persists in a surface drinking-water supply for 30-100 d and that someone drinks 2 L/day from this supply, we get an estimated contact with the deposited cadmium of 60-200 L for a single event. For that person, the potential dose for a single event is

$$\text{potential dose} \approx C_{water} \times (60\text{-}600)\ L$$
$$= (0.24\text{-}0.8) \times (C_{air} \times ET)\ \mu g\ [\text{over} \approx 30\text{-}100\ d].$$

That value is comparable with and slightly lower than the direct exposure to contaminated air by inhalation during the exposure event. However, there is greater uncertainty associated with this value than with the direct inhalation-dose estimate. This pathway is relevant only in cities like Fort Wayne, IN, where a public water supply was exposed to the same degree of air concentration as the population. In most of the cities under consideration, that was not the case.

REFERENCES

Allott, R.W., M. Kelly, and C.N. Hewitt. 1992. Behavior of urban dust contaminated by chernobyl fallout: Environmental half–lives and transfer coefficients. Environ. Sci. Technol. 26:2142--2147.

ATSDR (Agency for Toxic Substances and Disease Registry). 1993. Toxicological Profile for Cadmium. TP-92/06. Atlanta, Ga.: Agency for Toxic Substances and Disease Registry.

Baes, C.F. III, R.D. Sharp, A.L. Sjoreen, and R.W. Shor. 1984. A Review and Analysis of Parameters for Assessing Transport of Environmentally Released Radionuclides Through Agriculture. ORNL–5786/NTIS DE85-000287. Oak Ridge National Laboratory, Oak Ridge, Tenn.

Bodek, I., W.J. Lyman, W.F. Reehl, and D.H. Rosenblatt, eds. 1988. Environmental Inorganic Chemistry: Properties, Processes, and Estimation Methods. New York: Pergamon.

Bowen, H.J.M. 1979. Environmental Chemistry of the Elements. New York: Academic.

Davidson, C.I., W.D. Goold, T.P. Mathison, G.B. Wiersma, K.W. Brown, and M.T. Reilly. 1985. Airborne trace elements in Great Smoky Mountains, Olympic, and Glacier National Parks. Environ. Sci. Technol. 19(1):27-35.

Duggan, R.E., and P.E. Corneliussen. 1972. Dietary intake of pesticide chemicals in the United States (III), June 1968–April 1970. Pestic. Monit. J. 5:331-341.

Eggleton, A.E.J., and N. Thompson. 1961. Loss of fluorescent particles in atmospheric diffusion experiments by comparison with radioxenon tracer. Nature 192:935-936.

Elinder, C.-G. 1985. Cadmium: Uses, occurrence, and intake. In Cadmium and Health: A Toxicological and Epidemiological Appraisal. Vol. I. Exposure, Dose, and Metabolism. Effects and response. L. Friberg, C.-G. Elinder, T. Kjellström, G.F. Nordberg, eds. Boca Raton, Fla.: CRC Press.

EPA (U.S. Environmental Protection Agency). 1981. Health assessment document for cadmium. EPA-600/8-81-023. Research Triangle Park, N.C.: U.S. Environmental Protection Agency, Environmental Criteria and Assessment Office.

EPA (U.S. Environmental Protection Agency). 1992. Dermal Exposure Assessment: Principles and Applications EPA/600/8-91/011B. Office of Health and Environmental Assessment, U.S. Environmental Protection Agency, Washington, D.C.

Friberg, L., Piscator, M., Nordberg, G. F., and others. 1974. Cadmium in the Environment, 2nd Ed. Boca Raton, Fla.: CRC.

Gartrell, M.J., J.C. Craun, D.S. Podrebarac, and E.L. Gunderson. 1986. Pesticides, selected elements, and other chemicals in adult total diet samples, October 1980–March 1982. J. Assoc. Off. Anal. Chem. 69:146-159.

Gunderson, E.L. 1995. Dietary intakes of pesticides, selected elements, and other chemicals: FDA total diet study, June 1984–April 1986. J. AOAC Int. 78(4):910.

IARC (International Agency for Research on Cancer). 1993. Beryllium, Cadmium, Mercury, and Exposures in the Glass Manufacturing Industry. IARC Monographs on the Evaluation of Carcinogenic Risks to Humans, Vol. 58. Lyon, France: International Agency for Research on Cancer.

LaGoy, P. K. 1987. Estimated soil ingestion rates for use in risk assessment. Risk Anal. 7:355-359.

Leighton, P.A., W.A. Perkins, S.W. Grinnell, and F.A. Webster. 1965. The Fluorescent particle atmospheric tracer technique. J. Appl. Meteorol. 4:334-335.

Mahaffey, K.R., P.E. Corneliussen, C.F. Jelinek, and J.A. Fiorino. 1975. Heavy metal exposure from foods. Environ. Health Prespect. 12:63-69.

McKone, T.E., and P.B. Ryan. 1989. Human exposures to chemicals through food chains: An uncertainty analysis. Environ. Sci. Technol. 23:1154-1163.

McKone, T.E., and J.I. Daniels. 1991. Estimating human exposure through multiple pathways from air, water, and soil. Regul. Toxicol. Pharmacol. 13:36-61.

Mertz, W. 1981. The essential trace elements. Science 213:1332–1338.

Murthy, G.K., U. Rhea, and J.T. Peeler. 1971. Levels of antimony, cadmium, chromium, cobalt, manganese, and zinc in institutional total diets. Environ. Sci. Technol. 5:436.

Nazaroff, W.W, and G.R. Cass. 1989. Mathematical modeling of indoor aerosol dynamics. Environ. Sci. Technol. 23:157-166.

Nriagu, J.O. 1980. Cadmium in the Environment. Part I: Ecological Cycling. New York: John Wiley & Sons.

Nriagu, J.O. 1981. Cadmium in the Environment. Part II: Health Effects. New York: John Wiley & Sons.

Saltzman, B.E., J. Cholak, and L.J. Schafer. 1985. Concentrations of six metals in the air of eight cities. Environ. Sci. Technol. 19:328-333.

Schroeder, H.A., and J.J. Balassa. 1961. Abnormal trace metals in man: Cadmium. J. Chron. Dis. 14:236-258.

Schroeder, W.H., M. Dobson, and D.M. Kane. 1987. Toxic trace elements associated with airborne particulate matter: A review. JAPCA 37:1267-1285.